DISCARDED BY

MACPHÁIDÍN LIBRARY

CUSHING-MARTIN LIBRARY
STONEHILL COLLEGE
NORTH EASTON, MASSACHUSETTS 02357

# DNA ON TRIAL

## Genetic Identification and Criminal Justice

Edited by

### Paul R. Billings, M.D., Ph.D.
*Vice Chairman, Department of Medicine*
*Chief, Division of Genetic Medicine*
*California Pacific Medical Center*

 **COLD SPRING HARBOR LABORATORY PRESS 1992**

# DNA ON TRIAL
## Genetic Identification and Criminal Justice

© 1992 by Cold Spring Harbor Laboratory Press
All rights reserved
Printed in the United States of America
Book design by Emily Harste
Cover design by Jim Suddaby

**Library of Congress Cataloging-in-Publication Data**

DNA on trial: genetic identification and criminal justice /
   edited by Paul R. Billings.
      p.   cm.
   Includes bibliographical references and index.
   ISBN 0-87969-379-7 (cloth)
        0-87969-399-1 (pbk.)
   1. DNA fingerprints.  2. Evidence, Expert.   I. Billings, Paul R.
RA1057.55.D65  1992
614'.1--dc20                                                    92-19236
                                                                                  CIP

The articles published in this book have not been peer-reviewed. They express their authors' views, which are not necessarily endorsed by the Banbury Center or Cold Spring Harbor Laboratory.

Authorization to photocopy items for internal or personal use, or the internal or personal use of specific clients, is granted by Cold Spring Harbor Laboratory Press for libraries and other users registered with the Copyright Clearance Center (CCC) Transactional Reporting Service, provided that the base fee of $3.00 per article is paid directly to CCC, 27 Congress St., Salem, MA 01970. [0-87969-379-7/92 $3 + 00]. This consent does not extend to other kinds of copying, such as copying for general distribution, for advertising or promotional purposes, for creating new collective works, or for resale.

All Cold Spring Harbor Laboratory Press publications may be ordered directly from Cold Spring Harbor Laboratory Press, 10 Skyline Drive, Plainview, New York 11803. Phone 1-800-843-4388 (in Continental U.S. and Canada). All other locations: (516) 349-1930. FAX: (516) 349-1946.

# Contents

Preface, v

Preface to Paperback Edition, vii

**Gene Technology: Views of its Criminal Justice Applications** /
Paul R. Billings     1

**Galton's Regret: Of Types and Individuals** / Paul Rabinow     5

**Reasons for Doubt: Legal Issues in the Use of DNA Identification Techniques** / Marjorie Maguire Shultz     19

*Addendum:* **An Analysis of Decisional Law Governing the Use of DNA Evidence** (As of January 1992) / David A. Gass and Marjorie Maguire Shultz     43

**Forensic DNA in the Trial Court 1990–1992: A Brief History** / Jeffrey Baird     61

**Reliability of Statistical Estimates in Forensic DNA Typing** / Bruce Budowle, Keith L. Monson, and James R. Wooley     79

**Statistical Issues in DNA Identification** / Donald A. Berry     91

**Public Policy for Forensic DNA Analysis: The Model of New York State** / Jeroo S. Kotval     109

**The Impact of DNA-based Identification Systems on Civil Liberties** / Philip L. Bereano     119

**Genetics, Race, and Crime: Recurring Seduction to a False Precision** / Troy Duster     129

**DNA Data Banking and the Public Interest** / Nachama L. Wilker, Steven Stawski, Richard Lewontin, and Paul R. Billings     141

Index, 151

# Preface

This volume was created following a symposium held at the 1991 meeting of the American Association for the Advancement of Science. Nachama Wilker and the members of the Human Genetics Committee of the Council for Responsible Genetics cosponsored the event and were instrumental in its creation. Mia Fuller provided important help with many aspects of this project. In addition, Paul Rabinow, Troy Duster, Charles Cantor, Keith Marton, and the California Pacific Medical Center have provided intellectual support and free time to complete this work. Robin Fox was exceptionally generous with time and insight into the legal issues explored in this volume. Finally, Vicki Fischer in San Francisco and Mary Cozza and Patricia Barker at Cold Spring Harbor Laboratory provided consistently considerate and superb technical support.

Two other individuals must also be mentioned. Both Ruth Hubbard and Jon Beckwith are examples of the best that academia has to offer. As committed professionals and good citizens, they have been role models to many students. As mentors and available friends, they have been of great personal value to me. This book would not have been possible without them.

**P.R. Billings**

# Preface to Paperback Edition

The political philosopher Hannah Arendt has noted that "Justice, but not mercy, is a matter of judgment..."[1] A trial is a process for examining evidence in order to establish facts that can inform and facilitate a judgment about the commission of a crime, and the guilt or innocence of an individual with respect to such an event. The goal is justice. Such a process and its goal comprise a complicated and often uncertain endeavor.

This book is about a method of individual identification using a type of analysis of human DNA. Identification is often a key fact to be established at trial. During investigations that precede trials, this technique may be used to exclude suspected perpetrators. In the chapters that follow, the DNA-based method is scrutinized from many points of view. Issues surrounding scientific findings and the transmission and application of scientific data in the context of a courtroom are raised. An intricate controversy is illustrated, complicating any simple notion of identification. How this complex issue is perceived and understood is relevant to both judgment and justice.

Two recent examples of apparent misperceptions of the role of DNA analysis have impressed me. Reporting on the identification of a man accused of rape, a newspaper quoted a crime laboratory source as stating that the probability that the DNA sample analyzed was from a person other than the accused was about 13 billion to 1.[2] Yet the article did not mention that the critical issue, given a "match" between crime scene sample and that taken from the accused, is the comparison between that individual's DNA type and the 3 billion or so other possible sources of human DNA. The DNA identification process should posit only two a priori possibilities: This DNA is from a unique individual when the probability of a match being equivalent to identification equals 3 billion to 1, or that these data could have come from 2 or more of the 3 billion human beings, in which case the probability of a match being equivalent to identification is less than 3 billion to 1.

In addition, the story did not discuss the fact that the accused individual was in jail in another state, an issue that may be relevant to the reliability of the DNA typing result as a means of identification.[3] It also did not note that the National Academy of Science panel which reviewed the forensic use of DNA identification had specifically concluded that calculations leading to enormous odds like 13 billion to 1 were inappropriate.[4] This newspaper account thus made simple a situation which on closer examination was not.

A second recent publication concerned the so-called "prosecutor's fallacy" — the belief that the statistics surrounding DNA identification are clear and understandable.⁵ A case is described in which the prosecutor and judge seemingly confused data about the "match" probability and its relationship to the identification of a criminal, potentially confusing the jury. Prosecutors, judges, and defense attorneys may all understand the same identification information differently. However, judges and juries must make just judgments. Their awareness of the ambiguities in the data presented to them ought to be important in the process in which they are engaged. The tenet that good science leads reliably to more questions, and less so to truth, may extend to venues outside the experimental laboratory.

I devoted time and effort to producing this book because I thought it might illuminate the first example of the widespread application of a DNA-based method, for a nonmedical purpose, in an important social setting. I hypothesized that the clarity and elegance of the theory and techniques of DNA analysis, like the reductionism inherent in much of human genetic data, would not transfer simply from the laboratory to other social contexts. I believe that the responses to this book's publication, and the continuing discussion of the issues that it raises, confirm my supposition. In addition, as a citizen, I have learned something of value about judgment and justice. I thank Roxanne Mykitiuk, Michael Yesley, and Robin Fox for their comments.

*May, 1994*  P.R. Billings[*]

[1]Arendt, H. 1984. *Eichmann in Jerusalem — A report on the banality of evil.* Penguin Books, N.Y.
[2]Chapman, G. 1994. Inmate linked to murders. *Oakland Tribune* Jan. 11, p. 1.
[3]Berry, D.A. 1992. This volume, pp. 91–96.
[4]National Academy of Sciences. 1992. DNA technology and forensic science. In *Committee on DNA Technology in Forensic Science Report.* National Academy Press, Washington, D.C.
[5]News and Views 1994. *Nature* **367**:101–102.

[*]Present address: Associate Clinical Professor of Medicine, Stanford University and Acting Chief, General Internal Medicine, Palo Alto Veterans Affairs Medical Center, Palo Alto, CA 94304.

# DNA ON TRIAL

To my mother

Marta Amelia Seligman Palmer Billings Salinger, M.D.

# Gene Technology: Views of its Criminal Justice Applications

**PAUL R. BILLINGS**

Department of Medicine
Pacific Campus, California Pacific Medical Center
San Francisco, California 94115
and University of California
San Francisco, California 94143

In a recent article about DNA identification systems, a juror commented on the biological evidence presented following the conviction of the defendant, "You can't argue with science." (Neufeld and Colman 1990). Indeed, forensic science data and expert testimony delivered by scientists and physicians can be influential in the judicial process. Unfortunately, juries and the public from which they are drawn have an incomplete understanding of scientific controversy and the limitations of, and ambiguities surrounding, scientific truth. For instance, what is proven about the use of DNA techniques for identification purposes and what, in contrast, are initial findings in need of further scientific scrutiny may not be clear even to the *cognoscenti* of science. Thus, although improvements in general scientific education might make the transition of new technologies into social settings like courtrooms easier, there will always be conflicts, controversies, and misunderstandings as powerful methods are applied within differing social contexts.

This point could not have been more graphically made than by a series of stories that appeared on consecutive days on the front page of *The New York Times*. The first story summarized the findings of a near-final draft of a report by the National Academy of Science (NAS) on DNA identification methods. It reported the conclusion that DNA identification systems were not adequately studied to be used safely in forensic settings (Kolata 1992a).

The next day, a news conference was conducted and attended by the chairman of the NAS panel, the respected human geneticist Victor McKusick. A story indicating that DNA techniques could be used in courtrooms was published (Kolata 1992b). No novel scientific facts relevant to this question had been uncovered in the 24 hours separating these two news stories.

Finally, on the third day, a summary report of the previous stories appeared. It acknowledged the ambiguities, conflicts, and politics that color the movement of biological methods from laboratory to courtroom (Kolata 1992c). Subsequent reports concerning conflict of interest allegations and coercive practices employed by the Federal Bureau of Investigation (FBI) in several cases where DNA methods of identification were used only emphasize the *realpolitik* encountered when biology meets the world of crime and punishment in our country (Anderson 1992a,b; Roberts 1992).

DNA has associated "genetic mythologies": of limitless biological relevance, of being the most basic of biological information, of genetic determinism, and of "high tech" scientific truth. These myths influence the perceptions and acceptance of genetic information and methods by our society. The chapters that follow illustrate how DNA information is viewed from differing perspectives as it enters the courtroom. Given the type of information, its historical (and mythic) associations, and the criminal justice context, many of the findings noted may be unique; on the other hand, there are probably paradigms within this analysis for the wide-ranging applications of new biotechnologies that seem likely to occur.

DNA identification systems have rapidly become applied in scientific investigation (to track biological samples), criminal and civil proceedings, and searches for missing people. For instance, babies misplaced after birth and families separated during political repressions in Argentina have been assessed using DNA identification methods (King 1991). A measure of the impact of this new technology is that the NAS and the Office of Technology Assessment, a research office of the United States Congress, have already issued controversial review reports on the use of these techniques (Nishimi 1990; National Research Council 1992).

The use of fingerprinting for identification purposes was developed early in this century and has gained widespread acceptance. After its introduction, 20 years passed before the technique had been widely studied and validated, at least enough to gain general judicial acceptance. We now know that the fingertip pattern is a polygenic trait and may contain other information which is sometimes useful for medical diagnosis. Although it is not generally accepted in this country, the study of dermatoglyphics is widely used in Europe (Schaumann and Alter 1976). This national and cultural difference is relevant to current questions about DNA-based systems; it is claimed in this country (without evidence) that no information aside from the unique electrophoretic pattern is contained in currently employed DNA identification methods. New evidence about the biology of highly repetitive DNA sequences raises questions about this assumption (see below).

To date, thousands of investigations and cases in both criminal and civil proceedings have used DNA identification techniques. Because these processes were only discovered in the 1980s, experience with these methods has been much less than that required for acceptance of the earlier fingertip identification systems. When a scientific discovery is sped into application, inadequate time may be allowed for the scientific community to scrutinize the techniques, for the peer review process to eliminate spurious information, and for controversies to be resolved. Although many forces have brought DNA identification systems quickly to the courtroom or marketplace, a significant influence must surely have been the profit motives of biotechnology companies who conduct these tests, as well as our abhorrence of violent criminal activity which they are purported to deter.

Many of the principles underlying the currently available DNA identification tests are not substantially different from those that evolved from information about blood typing or typing of the HLA antigen system. The use of highly variable regions of DNA or "minisatellites" was initially reported by Alec Jeffreys (Jeffreys et al. 1985). His techniques have been widely exploited in the United Kingdom and are used in a variety of social processes, including criminal and immigration investigations. The advantages of DNA identification systems are that they can be done on small amounts of material, the material can be very old (even from mummies), and, in principle, the technique is highly specific.

Many interesting scientific questions remain unanswered about these methods. They include why such highly variable regions of human DNA exist, if they have any biological significance, and whether their location is in some cases near important genes for normal processes or near genes related to illness. The recent finding that short, highly repetitive sequences of DNA reside near, and influence, clinically important genes only emphasizes how little is known about human DNA or its ultimate biological importance (Richards and Sutherland 1992).

Dr. Jeffreys (pers. comm.) has reported that through the use of his identification techniques, 6% of identical twins were found to have nonidentical patterns. This surprising finding, if it can be reproduced, would suggest that identical twins are not identical, that identical twins become less identical over time, and/or that technical errors inherent in these methods are more common than is presently appreciated.

Of course, we do know much more about human DNA than ever before. Analyses of long stretches of human DNA sequence indicate that an individual's DNA varies at every 100 to 1000 basic DNA units (bases). Given that the total content of DNA in a single cell is approximately 6 billion bases, this suggests that each of us varies, individual to individual, by at least 6 million bases. However, looking at the same data in another way, 99–99.9% of our DNA is identical from one individual to the next. This fact, along with others in the currently used technical analyses of DNA, produces a tendency toward "sameness" when available methods are applied. Present DNA identification systems, and those in development, are directed at isolating and analyzing that 1–0.1% of our DNA which is unique. Theoretically, the most reliable method for detecting the DNA differences between human samples would depend on actual sequencing of known variable regions of human DNA. Many of the concerns raised about current techniques will be answered when such an approach is technically feasible and affordable.

In the chapters that follow, the authors explore how successful the attempt has been to reproducibly identify unique differences in DNA for forensic and other social purposes. They discuss how these methods have been applied in the real, nonrational world of criminals and prosecutors, expert scientists and physicians, biotechnology companies, and a public at times bewildered and woefully undereducated about genetics. It is also noted how this new technology challenges important principles of privacy, civil rights, and ethics. It can be concluded that we are quickly approaching a crisis in the use of these methods; the definitive question is being posed—*Does the application of DNA methods for personal identification by governments or private institutions contribute to the public welfare, violate important civil rights, and have an appropriate social mandate?*

**REFERENCES**

Anderson, C. 1992a. Conflict concerns disrupt panels, cloud testimony. *Nature* **355**: 753.
———. 1992b. FBI attaches strings to its DNA database. *Nature* **357**: 618.
Jeffreys, A.J., V. Wilson, and S.L. Thein. 1985. Individual-specific "fingerprints" of human DNA. *Nature* **316**: 76.
King, M.C. 1991. An application of DNA sequencing to a human rights problem. In *Molecular genetic medicine* (ed. T. Friedman), p. 117. Academic Press, San Diego.
Kolata, G. 1992a. U.S. panel seeking restriction on use of DNA in courts. *The New York Times*, April 14, p. A1.
———. 1992b. Chief says panel backs courts' use of a genetic test. *The New York Times*, April 15, p. A19.
———. 1992c. DNA fingerprinting: Built-in conflict. *The New York Times*, April 17, p. A13.
National Research Council. 1992. DNA technology and forensic science. National Academy Press, Washington, D.C.
Neufeld, P.J. and N. Colman. 1990. When science takes the witness stand. *Sci. Am.* **262**: 46.
Nishimi, R.Y. 1990. Genetic witness: Forensic use of DNA tests. Office of Technology Assessment (OTA). U.S. Government Printing Office (#052-003-012-031), Washington, D.C.
Richards, R. and G. Sutherland. 1992. Dynamic mutations: A new class of mutations causing human disease. *Cell* **70**: 709.
Roberts, L. 1992. Science in court: A culture clash. *Science* **257**: 732.
Schaumann, B. and M. Alter. 1976. *Dermatoglyphics in medical disorders.* Springer-Verlag, New York.

# Galton's Regret:
# Of Types and Individuals

**PAUL RABINOW**
Department of Anthropology
University of California
Berkeley, California 94720

> *We read of the dead body of Jezebel being devoured by the dogs of Jezreel, so that no man might say, "This is Jezebel," and that the dogs left only her skull, the palms of her hands, and the soles of her feet; but the palms of the hands and the soles of the feet are the very remains by which a corpse might be most surely identified, if impressions of them, made during life, were available.*
>
> Sir Francis Galton

The basic argument of this paper is that the principal technical means of individual identification—fingerprinting—is based on a separation of individuals and populations. Sir Francis Galton, the founder of modern eugenics and the systematizer of fingerprint methods, regretted this separation. Emerging methods of identification based on DNA proceed in the opposite direction; they currently can identify individuals only on the basis of population genetics. DNA fingerprinting is really DNA typing. The main critics of the forensic use of this technique have rightly called for more testing, more standardization, more quality control, and above all, better population statistics to ensure more accurate identification of individuals. The question is, What will happen when all these improvements are brought about? The answer is a re-legitimization of the use of biological classifications (race, ethnic groups) in a highly charged social arena. This paper concludes by wondering whether the scientists who rigorously criticize the current DNA technology shouldn't be worried by the potential consequences of legitimatizing racial categories in a society such as exists in the United States. Rather than adopting a negative stance toward the new technologies, perhaps research should be directed at finding individualizing markers not based on population differences.

## FINGERPRINTS

Sir Francis Galton (1892) began his book on fingerprinting with the first of a long series of Victorian commonplaces: how modern science has distinguished truth from appearance. Of the two kinds of marks on the hands and soles of the feet, one offered false hopes and the other true clues. The creases of the palm, to which a great deal of significance had been attributed over the centuries and across cultures, revealed no real secrets. Their patterns literally embodied only the sheerest physical index of use, labor or leisure. Palm prints, for Galton, did indicate something about the individual's experience and station in life but not enough to provide a means of identification to sort individual from individual, only class from class. On the other hand, the so-called papillary ridges were the most crucial anthropological data. These ridges, Galton concluded, offered the surest source of individual identification precisely because they were, for reasons unknown, distinctive to each person and remained identical throughout life. "There is no prejudice to be overcome in procuring these most trustworthy sign-manuals, no vanity to be pacified, no untruths to be guarded against" (Galton 1892). Although they revealed nothing about experience, they were indelible marks of individuality.

According to Galton, the first scientific documentation, as opposed to the unscientific use in many cultures, of the systematicity and hence usefulness of fingerprints for identification purposes was carried out in 1823 by Dr. Purkenje of the University of Breslau. The first practical usage of fingerprinting took place in Bengal. As a Major Ferris of the India Staff Corps put it, "The uniformity in the colour of the hair, eyes and complexion of the Indian races renders identification far from easy" (Galton 1892). The "proverbial prevalence of unveracity" of the Oriental races provided the motivation for these gentlemen to perfect a reliable identification system, one whose basis lay in a marker beyond or below the cunning will of native or criminal. These colonial officers had stumbled across a "sign-manual that differentiates the person who made it, throughout the whole of his life, from all the rest of mankind" (Galton 1892). Physiological research demonstrated that a distinctive pattern developed as early as the fourth month of pregnancy and was fully formed by the sixth month. Although through use or injury or growth the shape of the fingertip did change, the number of ridges and their minutiae, which compose the pattern, did not. Galton demonstrated that identification could usually be shown with just five elements. Once a system was set in place to photograph the prints acquired through simple inking methods, little skill was necessary to obtain fingerprints and not much more to interpret them.

During the 1880s, the reigning system of criminal identification was in a positivistic sense much more scientific. Alphonse Bertillon had developed a system of 12 anthropomorphic measures (head length, head breadth, middle finger length, foot length, length and breadth of the ear, height of the bust, eye color, etc.; small, medium, or large). These measurements yielded almost a million possible combinations, easily increased to infinity with the addition of a few more measurements when necessary (Bertillon 1881; Bertillon and Chervin 1909). "Bertillonage" was performed by three operators and three clerks during an examination taking 6–8 minutes. Data were entered via a code on cards, which were then photographed, making it easy to reproduce them for distribution to other identification centers (courts and prisons). The French state gave Bertillon's system official status in 1883. Galton had gone to France to witness the use of Bertillonage directly and was impressed with the analytic methods as well as their institutional implementation.

They were scientific and easily obtained. In his book, Galton acknowledged the comprehensive scope of Bertillon's measurements but insisted that fingerprinting was more certain. "Bertillonage can rarely supply more than grounds for very strong suspicion: the method of finger prints affords certainty...Let us not forget two great and peculiar merits of finger prints: they are self-signatures, free from all possibility of faults in observation or of clerical error; and they apply throughout life" (Galton 1892). The same could not be said for Bertillon's measurements. This paper is not the place to rehearse the decline and fall of nineteenth-century physical anthropology's dependence on phenotypic measurements (Gould 1981; Nye 1984; Wechsler 1982). I introduce Bertillon's system because historically it was replaced by Galton's and because many of its analytic strategies, thoroughly transformed a century later through the rise of an entirely different genetics and molecular biological techniques (i.e., measurement, distribution, comparison, and statistical evaluation), are the subject matter of the following sections.

Sir Francis Galton was frankly a little disappointed with his studies on fingerprints. Although fingerprints did provide a powerful device for the identification of criminals (and everyone else for that matter), they revealed nothing about individual character or group affiliation. After examining prints from English, Welsh, Jews, Negroes, and Basques, Galton bowed to his results; there was an identical range and frequency of fingerprint elements and types. Analysis of the prints of artists, scientists, and idiots revealed no systematic differences. Galton admitted that he had had "great expectations, that have been falsified, namely, their use in indicating Race and Temperament" (Galton 1892). He was forced to conclude (not without a certain regret), "Consequently genera and species are here seen to be formed without the slightest aid from either Natural or Sexual Selection" (Galton 1892). Fingerprints were individual, yet bore no trace of character, society, or evolution, and to that extent, constituted for the Victorian founder of eugenics a disappointment.

Galton's method was revised in 1897 by another Englishman, and that system is still in use worldwide today (Fincher 1989). Obviously, the story of fingerprinting's triumph over rival identification methods has a more complicated institutional and jurisdictional history, which remains to be written. During the 1970s and 1980s, laser technology has enabled forensics experts to make increasingly precise prints from minimal traces; vastly more powerful computers have made the system more efficient by allowing rapid comparison searches, which would have taken years by earlier technology. Regional computer networks have been put in place. The basic principles of fingerprinting, however, have not changed—Galton's regret remains.

## THE FRYE TEST

Considering how important science and technology have been in America since its inception, the legal precedents upon which the admissibility of scientific expertise have rested in the American legal system are surprisingly tardy and thin. As opposed to patent law, which was laid down in the early days of the republic and continues to inform the legal context in which contemporary inventions such as recombinant DNA methods and processes are adjudicated, it was only in 1923 that general legal principles were formulated for the admissibility of scientific evidence. This judicial threshold was crossed in the Frye decision, in which it was held that a precursor of the polygraph test was not admissible as evidence in a murder trial. Of more lasting importance than the specific decision itself were the standards that the court proposed for admitting scientific expertise into the courtroom. The Court of Appeals for the District of Columbia held, "Just when a scientific principle or discovery crosses the line between the experimental and demonstrable stages is difficult to define. Somewhere in this twilight zone the evidential force of principle must be recognized and while courts will go a long way in admitting expert testimony deduced from a well recognized scientific principle or discovery, the thing from which the deduction is made must be sufficiently established to have gained general acceptance in the particular field in which it belongs" (Neufeld 1989). Although the court cited no precedent, its own decision established the precedent to which contemporary debate still defers.

The court's decision raised a number of separate issues bearing on how truth is articulated in the legal context that remained basically uncontested for decades. They began to be challenged only in the mid-1960s. The points at issue are (1) identifying the appropriate scientific field or fields; (2) quantifying "general acceptance" in the particular field; (3) deciding whether proof of validity of "the thing from which the deduction is made" must support the underlying scientific theory or the technique or perhaps both. These criteria seemed to have received little or no criticism, either legal, technical, or cultural, in the decades following the Frye decision. Fingerprinting, as well as improved polygraphs and other such devices, was used widely by local and national law enforcement agencies and routinely accepted into evidence by the courts. Because this broad acceptance must have had echoes and anchors in the broader social arena, it would be illuminating to trace the development of the popular image of the police laboratory during the interwar and immediate postwar periods. These were decades, it seems, in which the law, science, and the public representations of the truth were made to harmonize.

An example of the reexamination and devastating criticism of previously unchallenged truths can be seen in the diphenylamine or "paraffin" test used to detect gunshot residue on the hands. The procedure was introduced in the early 1930s and admitted as valid by the courts in 1936, even though an official FBI evaluation of 1935 questioned its authoritativeness. It was not until 1967 that a comprehensive independent study evaluated the evidence and showed the test's unreliability. "Among the substances that also gave a positive test result and could easily be found in residue form on anyone's hand were evaporated urine, tobacco ash, fertilizer, various pharmaceuticals, and colored finger nail polish" (Neufeld

1989). The paraffin test had met the Frye standard of "general acceptability" for 30 years and had been used widely. It was not until the mid-1970s that a federal policy on evidence was formulated. Most surprising of all, it was not until the late 1980s with the introduction of so-called DNA fingerprinting that sustained debate about how unclear the standards of scientific expertise really are unfolded.

As did so many other aspects of American society and culture, forensic institutions and procedures for establishing truth began to change in the mid-1960s, one of those periodic conjunctures of accelerated modernization of law, technology, and public opinion in the United States. The institution that embodies this modernization is the Law Enforcement Assistance Administration (LEAA), created in 1968 as part of the Omnibus Crime Control and Safe Streets Act. Supreme Court decisions strengthening defendants' rights explicitly hastened the search for technical and scientific evidence. As a Florida Appeals Court judge put it, "In this day and age. . . where recent decisions of the Supreme Court establish stringent guidelines in the investigative, custodial and prosecutorial areas a premium is placed upon the development and use of scientific methods of crime detection" (Gianelli 1980). The authority of both FBI and local police in general, as well as their taken-for-granted procedures of obtaining evidence and using it, was under attack. Although the long-range conservative response to the situation was to change those interpreting the law, in the short run, technological responses were accelerated as a means of establishing more convincing types of evidence, especially evidence not dependent on confession. As early as 1966, the California Supreme Court drew an important Fifth Amendment distinction between communicative or testimonial evidence, which is subject to privilege against self-incrimination, and physical or real evidence, which is not protected. The court noted that the privilege "offers no protection against compulsion to submit to fingerprinting, photographing, or measurements, to write or speak for identification, to appear in court, to stand, to assume a stance, to walk, or to make a particular gesture" (OTA 1990).

The Crime Laboratory Proficiency Testing Program initiated by LEAA concluded that many laboratories were below par. The response to this amateurism was unambiguous: Update and expand. More than half of the nation's crime laboratories came into existence after 1970, many through funding from the LEAA. Its presence accelerated the search for new technologies in the criminology area, including work on bloodstains, trace metal detection, and voice detection (Gianelli 1980). The modernization of forensics followed the analysis of the sociological literature on professionalization extremely well. As expected, one saw the differentiation of specialists and technicians, the formalization and jargonization of standards and practices, government regulation, and the like.

The first major reevaluation of the Frye principles came in challenges to voice print technology, a new device developed by police crime laboratory technicians in Michigan. This technology imitated fingerprinting with its ability to identify a person on the basis of basically unalterable phenotypical characteristics. Just as each individual has a unique set of fingerprints, so too, perhaps each individual has a unique voice. A technique called "sound spectroscopy" was developed, which produced an abstract printed pattern—a voice print—from a tape recording. When the scientific basis of the technique was challenged in court, the first question to be answered, following the Frye principles, was, since there was no preexisting scientific voice print research community, who were the appropriate scientific experts? Were they the inventors of sound spectroscopy, the Michigan State Police? Or did the identification of voice patterns fall into a wider series of specialties including anatomy, physiology, physics, psychology, and linguistics? Not only did the inventors of the technique appear to have a conflict of interests, but they were mere laboratory technicians, not scientists.

Under the auspices of LEAA, a national commission was formed to resolve the issue. The main point to be decided was simple: Were there scientific grounds to believe that speech contained a hidden pattern which could be used to identify a particular individual and, if so, did the sound spectrography technique adequately capture this particularity? Although the experts all agreed that there were regularities, no definitive elements or patterns could be identified; there was simply too much variation both between individuals and within an individual's own speech patterns. Again in typical reformist bureaucratic fashion, the committee did not close the case but called for a program of research and development leading to a science-based technology of voice identification. Although itself in many ways a technological throwback to an earlier era, voice printing nonetheless raised a series of interpretive issues arising from the Frye precedents. Who is the community of experts? What constitutes independent verification? What is the role of national agencies? These institutional issues remain open in contemporary debates about DNA methods: What should the relationship be between the businesses who perform the tests, the FBI, the academic community? Should there be an impartial national commission to set standards? How reliable are laboratory procedures? Which studies need to be done?

Although these institutional issues demonstrate continuity in framing the legal and bureaucratic dimensions of establishing truth, I believe there was an important turning point reached in the voice print controversy. Embedded in the National Commission's deliberations was a partial response to Galton's regret. The commission opined about an evolving science of voice identification, but, in fact, there was no evolving science of voice identification because it was not a science. Historically, failure of reform movements with their dubious claims to science have frequently resulted in more funding to improve the science. The susceptibility to voluntary control is an old theme stemming from the older moralistic disciplines of the nineteenth century, echoes of which are still found in the Frye decision's statement that individuals have an involuntary propensity to tell the truth. Along with these older strata was something not exactly new in itself (population genetics after all had been in existence for close to a century) but whose conceptual migration into forensics would be catapulted on to center stage by developments in recombinant DNA technology, the necessary *linking of the individual to a population.*

The invention of electrophoretic bloodstain analysis, in the early 1980s, provides an example of the crossing of this new threshold. Unlike voice printing, electrophoretic bloodstain analysis was a technique developed for entirely different purposes and then adapted to forensic purposes. Electrophoretic bloodstain analysis entered into the court system as early as 1978, but the majority of precedent-setting decisions concerning its admissibility as evidence took place in the mid-l980s. The courts proceeded from the voice printing debate foregrounding questions of credentials and professionalization, weighing conflicting claims of authority over content, e.g., Do results have to appear in peer-reviewed scientific journals to count as fact? (Harmon 1989). This attention to authority, procedure, and precedent, although appropriate in the logic of legal proof, tended to obscure qualitative differences in these technologies from the older ones.

DNA typing differs qualitatively from "the evolving science of voice identification." This claim obviously does not imply that procedures based on valid principles should carry automatic acceptance in individual cases, because technical mistakes and simple human errors like mislabeling samples are made all the time. Furthermore, it is the nature of an adversarial legal system for partisan interpretations to be presented tactically. According to Peter J. Neufeld, as of 1989, electrophoretic bloodstain evidence had been introduced in approximately 500 cases in New York. The defense challenged expert witnesses only twice with other expert witnesses. In both instances, following Frye hearings, the trial courts decided to exclude the serology evidence. "What does this result say about the quality of justice in the other 498 cases?" (Neufeld 1989). It means unequal access to technology emerges as a critical issue in a period of technological change. Scientific illiteracy of the judiciary, legal profession, and the public will be best protected against through education and adversarial proceedings.

Vigilance is the price of liberty. Challenging premature applicability of techniques is an appropriate stance to adopt. However, the most penetrating criticism about particular applications, nonregulated laboratories, and the like will eventually strengthen the truth claims of many of these new DNA techniques, not weaken them. The FBI is carrying out blind testing, looking at the probes, verifying the allele frequency tables, checking the population samples for variation in locale, calling for regulation of laboratories and the establishment of regional databases, and the like.

## MODERNIZING STANDARDS: FBI AND POPULATION

The FBI entered the field of DNA technology in late 1984 with a collaborative project with the National Institutes of Health. By 1986, agents were visiting laboratories in both England and the United States to survey and learn cutting-edge technologies. Private industry, government, and the university cooperated eagerly. In July 1987, an evaluation was made that the technology held great potential, and a decision was taken to establish an FBI research team at Quantico, Virginia (Hicks 1989). The center was staffed by specialists versed in traditional blood grouping and protein testing methods. The University of Virginia cooperated by organizing a course in molecular genetics for the agents. The FBI sees itself playing a catalyst role for streamlining, improving, and making these technologies more reliable and cost-effective. To this end, it holds training sessions for state and local officials. These sessions include presentations by "key scientists from academia, private sector laboratories, other DNA research centers, and the international crime laboratory community" (Hicks 1989). A visiting scientist program at the bureau's Forensic Science Research and Training Center enables local forensic technicians from across the country to return home fully trained in the techniques and prepared to set up cutting-edge forensic laboratories.

At the heart of the FBI's efforts is the issue of standardization. Having decided that the technology works, the FBI understood that without relatively standardized procedures and (even more importantly) without the prospect of a normative database, the future impact of this technology was limited. As John W. Hicks told the assembled experts at the Banbury Center of Cold Spring Harbor Laboratory, "There must be uniformity within the crime laboratory community on the DNA test methods used so that the profiles developed can be effectively cataloged and compared. There have already been some discussions within the forensic community exploring the feasibility of DNA data banks and uniform test protocols" (Hicks 1989). This call for norms involves more than cost-effectiveness or state control, although features of each are present. The need for norms derives from the nature of the project: Without an adequate database for the population genetics, identification of individuals could only be exclusionary at a relatively low level. As two other authors in the Banbury volume put it, "A number of organizations around the world are now considering development of large data bases containing DNA profiles of all individuals in specific populations. As the application of this technology continues to expand there is a growing concern as to whether standardization of systems is necessary" (Rose and Keith 1989). The answer is yes.

Of course, there will be some problems to iron out. Enforcing a common standard is not easy, particularly in a field in which private companies are doing the bulk of the analysis and want to retain proprietary rights over their techniques. In the general spirit of cooperation that currently marks these encounters, one participant suggested "an alternate approach that might satisfy the need for global compatibility could be the use of a completely standardized core system by all laboratories interested in interacting with a large data base, along with any additional systems the individual laboratories might choose to use" (Rose and Keith 1989). This approach would protect the proprietary rights of the companies involved while enriching the database from which probabilities are derived. The FBI advocates imitating the regionalization approach adopted by state officials running newborn screening programs. Not only does regionalization permit regulation of the procedures employed, it is also cost-effective and most easily meets the demands of the congressional oversight committee that blind screening be included as a control.

The FBI is explicitly seeking to reassure civil libertarians. The information included in the database would be chosen in such a manner that it could not be used abusively. Citing their choice of Jeffreys' variable number of tandem repeats (VNTR) approach, an FBI spokesman argues, "There is no known relationship between these numbers and any physical or mental condition. Selection of the genetic markers in the database will be made with an eye to eliminating the potential interest of the data to the private sector. Probes will not be used that are linked to disease conditions or personality traits" (Kearney 1990).[1] Although the genetic arbitrariness of the approach is meant to reassure, it does contain some ambiguities. The very reason the data are chosen—their supposed arbitrariness—opens up avenues for other arbitrary correlations, ones that might assign a meaning-function to them or interpret them as a marker of other conditions. Eric Lander cautions that we might well see random correlations being run: "This allele, at this locus about which I know nothing, tends to come up in rapists" (see comments by Lander in Ballantyne et al. 1989 [p. 12]). This is not an idle fantasy; we also read, "The justification for the development of these types of indices is based on the fact that individuals who commit violent crimes are often repeat offenders" (Kirby 1990) and "Several states are considering collecting DNA identification profiles of certain categories of convicted offenders" (Hicks 1989). Risk profile analysis, moving beyond socioeconomic variables, might well be hard to resist. The most obvious means of preventing such abuses would be to simply destroy the DNA upon which the tests were done; defense attorneys, however, might well resist this move because it denies them the possibility of verifying the analyses.

If risk correlations for their own sake are one danger, another is to find a more than correlational meaning to the data one has. Although the FBI seeks to be reassuring about the arbitrariness of the loci to be encoded, in an earlier part of the Banbury volume, Alec Jeffreys cautioned against assuming that "the regions most of us who are using this technology are looking at are noncoding. I am not aware of any formal proof that any of these regions is a noncoding region. There are examples showing that some of these regions are coding and that they are probably all coding" (see comments by Jeffreys in Ballantyne et al. 1989 [p. 36]). Jeffreys' point is a general and recurring one and carries with it a caution that might be phrased as follows: Although a great deal is being discovered, very little is yet understood. The effort to understand will no doubt follow the data being discovered.

---

[1]*Editor's note:* As mentioned in my introduction, the number of certain repetitive elements has recently been found to correlate with phenotypic features in several genetic disorders.

## WHY DNA FINGERPRINTING IS REALLY DNA TYPING

Eric Lander interrupted a forensic expert's summary, "[I]n forensic analysis, basically we compare the questioned sample with a known sample, side by side. It does not matter whether it is a voice screen, an infrared spectra, a GC chromatogram, or DNA; if they show identical patterns. . .", by asking, "Are they the same person?" "No," the expert responded, "the same pattern." Another expert added, "I do not think we are talking about unique identification. We are talking about things from profiles with some statistical evidence for a percentage of the population that may carry an array of types, and we are putting a value on them so that some people can properly evaluate what they mean" (see comments by Lander in Ballantyne et al. 1989 [p. 105]). DNA forensics is a question of *types*. The conceptual foundations of the typing methodology are uncontested, although the current technology will no doubt soon appear quaint.

With the population data, there are both factual and interpretation problems outstanding. The presence or absence of a polymorphism can exclude with absolute certainty. A suspect with one blood group cannot be responsible for a bloodstain from another blood group. However, inclusion depends on population genetics and, hence, on probabilities. Eric Lander points out that although there may be as many as three million sites of DNA variation between individuals (10% of the genome), only three or four restriction-fragment-length polymorphisms (RFLPs) have been used in forensics (Lander 1989). Grant that the various technical problems of proper laboratory work and overcoming degraded or insufficient samples have been solved, the main question mark for the RFLP approach to forensic identification is population genetics. The basic theory of population genetics has been established for a century now and its mathematics for half a century. However, neither the stakes nor the standards relevant to estimating the growth of a turtle population over centuries and the guilt or innocence of one human individual are yet commensurate.

Lander provides a succinct outline of the relevant principles. Assume that the DNA sample from the suspect matches that found at the crime, i.e., the RFLP lengths are the same on the radiograph. The question is, How likely is it that this match is a random one? The only way to answer the question is through population genetics. If we knew the distribution of the RFLP we would know the answer. Such data exist today only for very small samples (in almost all cases). It follows that the more loci one used, the higher the odds of a specific determination would be; therefore, multiply the probes. The concept is straightforward; empirically, there are often problems. First, has the correct population been identified? It was shown as early as 1918, during the First World War, that populations vary significantly by blood type. A great deal of work has been done since to detail this variation. The explanation for variation is genetic drift, e.g., Tay-Sachs disease is found with a greater frequency among East European Jews, a relatively isolated population.

Hypervariable loci are the best markers to use if one is seeking to differentiate populations. The study of this type of variation was modernized—standardized, operationalized, commercialized—by international studies of HLA genes, which control transplant rejection (among other factors). There is a large database for HLA variation. Variation is significant, e.g., the frequency of one HLA gene is 0.2% in Japanese and 19% in French Caucasians. Thus, a Frenchman would be 9025 times more likely to be a homozygote for HLA-A1 than a Japanese person would be. This range of variation has been found for other alleles. However, it is important to specify between which groups variation is being studied and that these groups are true groups in terms of population genetics criteria. Thus, do Puerto Ricans and Mexicans (Hispanic) really belong in the same category; do Russians and Italians (Caucasian)? Whatever the politics of classification was to establish this system of categories (Caucasian, black, Asian, American Indian, Hispanic), the categories are clearly too broad for accurate forensic population genetics.

A second problem is whether the sample is large enough that the observed frequencies accurately represent the true population frequencies. Statistical techniques exist to correct for nonrandom sampling, but they have not always been applied in forensics. More care in definition and selection and bigger samples must be introduced. Even with large numbers, it is not always easy to establish whether the sample is random or not. This is a serious problem. As Lander says, "Were it not unethical and unconstitutional, a court might compel randomly chosen individuals to provide blood samples in a population survey" (Lander 1989). He suggests the norm of standards used in political polling: a replicable, detailed, anonymous survey checked against census data.

A third problem is whether each locus is in Hardy-Weinberg equilibrium and the loci are together in linkage equilibrium: "This is nothing more than a restatement that each person's alleles must represent a random selection from the overall pool" (Lander 1989). This requirement is pertinent because it is the premise on which the calculations for randomness in a population depend. The HLA frequencies appear to be in Hardy-Weinberg equilibrium. "This observation is significant for the utility of the HLA-DQα marker in individual identification because it indicates that genotype frequencies can be reliably estimated from allele frequency data" (H. Erlich, unpubl.). The lack of Hardy-Weinberg equilibrium poses a problem for the VNTR data. Not surprisingly, "free interbreeding" appears not to be the American norm. We have ethnic and religious subgroups. Lander reiterates that results indicative of nonequilibrium do not invalidate the laws of population genetics and Mendelian inheritance. They reveal our inadequate understanding of the American population. This problem can be corrected through more studies and better statistical procedures. Another expert concurs that the issues will be resolved only "when the data bases are sufficiently expanded to provide large enough sample numbers to verify or reject the statistical approaches used" (Putterman 1990). The FBI is planning to undertake large sampling studies as a means of refining categories. In conclusion, Lander asks, "Do these precautions stand in the way of practical application of DNA typing technology to forensics? Fortunately, the answer is resoundingly negative" (Lander 1989). DNA typing is "revolutionary," "the ultimate identification scheme."

## CONCLUSION AND A MODEST PROPOSAL

To ensure adequate statistical protections, more finely drawn groups will have to be included in the database. Because it is currently in bad taste to refer directly to races or breeding populations, various subsets of this data pool will be called ethnic. Of course, these ethnic groups will be measured on a common grid—VNTRs or HLA frequencies—so that individuals can be placed in populations and subpopulations related to each other. At first, i.e., today, there will be criticisms articulated by high-ranking concerned scientists that there is too much lumping going on—Italians and Russians, Puerto Ricans and Mexicans. Appropriately, there will be finer and finer grids linking subethnic groups down to particular breeding populations, and no doubt a more sophisticated probability and statistics to do this. There will be scholarly and popular arguments about where the boundaries are and who constitutes them. It follows that although categories such as "race" or "ethnic group" may well continue in popular usage, they will begin to acquire a meaning that is more particularizing and relational than whether the surname is Spanish. These categories will be redefined—allelic polymorphism rates for HLA genes—and will then feed back into the broader cultural classifications with their political and social consequences in ways well worth monitoring.

While the FBI constructs regional "arbitrary" VNTR computer networks, other workers will map other more directly functional systems. The Human Genome Mapping Library at New Haven is the world's largest computerized repository of human gene mapping information. With the progress of the Human Genome Initiative, whose efforts will provide a common series of maps, these various data will eventually (in 5–10 years) merge. As one expert at the Cold Spring Harbor meeting put it, "With enough of these systems and variation in [noncoding VNTR] frequencies among populations, one could begin to look at it overall to infer racial origin." A colleague chimed in, "Even conventional genetic markers can occasionally give you precise racial data. For example, in a case last week, peptidase A 2-1 was found in evidence at the scene of the crime. It is highly likely that it originated from someone of Negroid origin" (Westin 1989).

The last 30 years have been a period of historic cultural redefinition of the traditional categories in the West, from race to gender to age, emphasizing plasticity. To mention only a few examples: thousands of "older people" now run in marathons, "sexual orientations" have multiplied. This emphasis on plasticity is not disappearing, but it is being challenged and supplemented in a variety of cultural domains by the reintroduction of supposedly biological constraints and givens: for example, the PMS syndrome or the homosexual hypothalamus. One can easily imagine gay senior citizen marathons. The power of the new genetics, biotechnology, and the umbrella of the Human Genome Initiative are providing "race" with a new legitimacy. However, the new techno-scientific understanding of population genetics will certainly be conflated with older cultural understandings of race, gender, and age. Some of the dangers inherent in this blurring are obvious; others are not. Once one has a database that meets all of Eric Lander's standards, should it be used to test affirmative action candidates to see if they "really" are Hispanic? American history is replete with older models of racial proof which could be drawn on. Of course, this trend is not restricted to the United States; some ethnic groups in China are demanding scientific proof of their (formerly culturally defined) minority status in order to be allowed to have more children.

The dangers of proceeding in the direction toward relegitimizing racial or "ethnic" biological categories in the forensic arena should be clear enough. However, simply opposing the use of the biotechnologies in these areas seems in equal parts futile and wrong. Perhaps we need to go back to Galton's regret. Perhaps some researchers should keep their data banks open for the possibility of looking for and discovering *individual* genetic variation, one not based in population genetics. Richard Lewontin argued in *Not In Our Genes* that differences among individuals were greater than among races in terms of IQ. IQ, he argued convincingly, was primarily environmental and individual (Lewontin 1984). What if we were to look for genes which distinguished individuals with the kind of specificity that fingerprints apparently do? Perhaps the genes for fingerprints—so individualizing—would be the place to start looking. Surely the techno-scientific imagination at the end of the twentieth century is capable of finding thousands of other distinctive alleles, ones not linked to race and temperament. Seek and ye shall find. Then Galton would continue to remain where he belongs—in purgatory.

## REFERENCES

Ballantyne, J., G. Sensabaugh, and J. Witkowski, eds. 1989. *DNA technology and forensic science*. Cold Spring Harbor Laboratory Press, Cold Spring Harbor, New York.

Bertillon, A. 1881. *Une application pratique de l'anthropometrie*. Paris.

Bertillon, A. and A. Chervin. 1909. *Anthropologie metrique, conseil pratiques aux missionnaires scientifiques sur la maniere de mesurer, de photographer et de decrire des sujets vivants et des pieces anatomiques*. Imprimerie Nationale, Paris.

Fincher, J. 1989. Lifting latents is now very much a high-tech matter. *Smithsonian* **20**: 216.

Galton, F. 1892. *Finger prints*. Macmillan, London.

Gianelli, P. 1980. The admissibility of novel scientific evidence: Frye v. United States, A half-century later. *Columbia Law Rev.* **80**: 1199.

Gould, S.J. 1981. *The mismeasure of man*. Norton, New York.

Harmon, R.P. 1989. The *Frye* test: Considerations for DNA identification techniques. *Banbury Rep.* **32**: 89.

Hicks, J.W. 1989. FBI Program for the forensic application of DNA technology. *Banbury Rep.* **32:** 209.

Jeffreys, A.J., Z. Wong, V. Wilson, I. Patel, R. Neumann, N. Royle, and J.A.L. Armour. 1989. Applications of multilocus and single-locus minisatellite DNA probes in forensic medicine. *Banbury Rep.* **32:** 283.

Kearney, J. 1990. The combined DNA index system (CODIS): A theoretical model. In *DNA fingerprinting: An introduction* (ed. L. Kirby), p. 284. Stockton Press, New York.

Kirby, L., ed. 1990. *DNA fingerprinting: An introduction.* Stockton Press, New York.

Lander, E. 1989. Population genetic considerations in the forensic use of DNA typing. *Banbury Rep.* **32:** 143.

Lewontin, R. 1984. *Not in our genes.* Pantheon Books, New York.

Neufeld, P.J. 1989. Admissibility of new or novel scientific evidence in criminal cases. *Banbury Rep.* **32:** 73.

Nye, R. 1984. *Crime, madness and politics in modern France. The medical concept of national decline.* Princeton University Press, New Jersey.

Office of Technology Assessment (OTA). 1990. *Genetic witness, forensic uses of DNA tests.* U.S. Government Printing Office, Washington, D.C.

Putterman, M. 1990. Probability and statistical analysis. In *DNA fingerprinting: An introduction* (ed. L. Kirby), p. 176. Stockton Press, New York.

Rose, S. and T. Keith. 1989. Standardization of systems: Essential or desirable? *Banbury Rep.* **32:** 319.

Wechsler, J. 1982. *A human comedy, physiognomy and caricature in 19th century Paris.* University of Chicago Press, Illinois.

Westin, A.F. A privacy analysis of the use of DNA techniques as evidence in courtroom proceedings. *Banbury Rep.* **32:** 25.

# Reasons for Doubt: Legal Issues in the Use of DNA Identification Techniques

**MARJORIE MAGUIRE SHULTZ**

Boalt Hall School of Law
University of California, Berkeley
Berkeley, California 94720

DNA identification is the latest in a long line of controversial forensic science methods (fingerprints, lie detectors, voiceprint). More than the others, however, this advance stimulates the imagination in ways akin to the mythical theft of fire from the gods. DNA-based techniques address growing fears about violent crime. They also respond to even more basic hungers; they seem capable of vanquishing the age-old foe of uncertainty itself.

These idealized hopes may explain why in the early wave of legal decisions, DNA identification techniques were readily accepted. Given that the *Frye* test, the law's traditional yardstick for determining admissibility of scientific evidence, directed attention to the general acceptance of a technique within its particular scientific field,[1] courts naturally embraced the new DNA evidence. Judges were understandably reluctant to reject the assistance of scientists testifying about a major breakthrough in the laboratory. Discoveries about DNA generated enormous prestige and visibility; experts willing to attack the methods were scarce.[2] Faced with complex presentations by highly funded biotechnology firms and without much in the way of their own expert assistance, the defense bar was largely unprepared to challenge this new type of evidence. Questions addressed to forensic use as opposed to theoretical validity of DNA methods initially went unrecognized in the rush to adopt methods that promised definitive identification or exculpation of suspected criminals.

More recently, court responses have become somewhat more divided. Although most courts are still admitting DNA evidence, growing awareness of problems in the forensic application of DNA identification has caused some courts to reject it as insufficiently validated.[3] This skepticism is not a product of generic know-nothing-

---

[1]Frye v. United States, 293 F. 1013 (D.C. Cir. 1923) (establishing standards for admissibility of novel scientific evidence, in this case systolic blood pressure deception test).

[2]Recently, allegations have been made that scientists questioning the use of DNA evidence have been intimidated and harassed. *See* Gina Kolata, *Critic of "Genetic Fingerprinting" Tests Tells of Pressure to Withdraw Paper*, N.Y. TIMES, Dec. 18, 1991, at A-20.

[3]*See, e.g.*, Decision on the Admissibility of DNA Identification Tests in People v. Castro, 1508/87 (N.Y. Sup. Ct. 1989) (evidence declaring a match declared inadmissible after prosecution and defense

ism or of categorical legal hostility to science, nor is it likely to represent any permanent bar. Rather, wariness about a premature or uncritical embrace of these powerful new techniques seems eminently appropriate in light of commentary here and elsewhere.

When science is imported into the courtroom, the leap from one context to another must be carefully analyzed. The two fields differ in significant ways. For purposes of assessing legal issues raised by DNA identification techniques, several themes of difference seem particularly relevant. First, science generally perceives itself as being preeminently about truth. In contrast, legal adjudication has truth as one goal, but sees other goals as equally or more important. Thus, trials, particularly criminal trials, seek truth, but they also express norms about the value of individuals, the role of government, the appropriate methods of truth-seeking, and the preferability of some risks over others. These value preferences are encoded in concrete legal institutions and procedures. Second, even to the degree that law, like science, does seek the truth, there are significant differences in the two fields' assumptions about the character and sources of truth. Whereas science envisions truth as a substantive quest, law tends toward procedural and functional definitions.

These differences in orientation and modes of knowing stem in part at least from a difference in role. Both science and law study the concrete and factual. However, science aspires primarily to identify and verify universal explanatory truths.[4] In law, in contrast, the central task is the resolution of particular disputes. Legal generalization plays a role in guiding and predicting solutions and in providing principled justificatory explanation for concrete outcomes. However, resolution of particular controversies remains not a means but an end in itself. Thus, it would be fair, albeit oversimplified, to say that science deals in particulars in order to determine generalizations; law deals in generalizations in order to determine partic-

---

experts together signed a declaration that the data were not significantly reliable enough to support an assertion of match or non-match). The case is discussed at length in Janet C. Hoeffel, *The Dark Side of DNA Profiling: Unreliable Scientific Evidence Meets the Criminal Defendant*, 42 STAN. L. REV. 465 (1990) (detailed critique of the potential for error in the laboratory and statistical analyses as used in the courtroom). See also United States v. Two Bulls, 918 F.2d 56 (8th Cir. 1990) (reversing conviction because trial court erred in admitting DNA identification evidence without determining if testing procedures in this case were conducted properly) vacated, rehearing en banc granted, 925F.2d1127 (1991); rehearing and indictment dismissed after death of defendant, 1991 U.S. App. LEXIS 6840; State v. Schwartz, 447 N.W.2d 422 (Minn. 1989) (DNA evidence admissible under a *Frye* test requiring inquiry into procedures in the specific case, and if full disclosure/access is provided to the defense; evidence in this case inadmissible for failure to comply with that standard); Commonwealth v. Curnin, 409 Mass. 218, 565 N.E.2d 440 (1991) (conviction reversed; DNA improperly admitted because of failure to establish general acceptance of the laboratory and statistical procedures used).

[4]An alternate and less well known philosophy of science is more modest and empiricist, seeking "local" rules rather than global truths.

ulars. The concrete and practical impact of legal decisions on particular individuals' lives reinforces the law's focus on values above and beyond simple truth and heightens its commitment to procedural and functional approaches to justice.

Other chapters in this volume amply demonstrate that issues about the technical and statistical, if not the theoretical, accuracy of DNA identification remain hotly debated within relevant scientific and technical communities. These disputes are, of course, implicated in any use of the procedures, including any legal uses. Beyond the problems raised by theoretical and technical accuracy alone, however, lie the additional and distinctive complications of policy, theory, and application that arise when identification techniques developed in scientific and commercial laboratories are transferred to adjudicative contexts. The factors that distinguish scientific modes from legal modes raise special problems for use of DNA identification techniques in legal fora. The primary focus of this chapter is on those special problems.

## BASIC LEGAL NORMS

### Individual Rights and Governmental Limits

In the American legal system, an interconnected web of values and assumptions defines the role of the individual, the role of the state, and the relationship between the two. Implicit in this web of principles is the conviction that the search for truth must coexist with other concerns. Perhaps the most central other value is the dignity and worth of the individual. The United States Constitution and the theories on which it is based accord the individual the highest priority. The state is seen as derivative, its function limited, especially by the rights of individuals.

Many settings and issues implicate these fundamental values of individualism and limited government. Principles such as nondiscrimination (treating likes alike) and privacy (protecting autonomous decision making, private spaces, and information), for example, express core concerns about the relationship between government and individuals. However, criminal prosecution is the primary locus where the individual directly confronts the power of the state. Commitments to truth and to those values that should coexist with truth are, therefore, deeply embedded in the institutions and procedures of criminal adjudication.

Although individuals have a considerable stake in the *protection* they can derive from truth, other values allow them to limit the ways in which the state can seek to find or use truth to *punish* them. Because the dignity of the individual must be respected, there are some things that state simply *cannot* do, no matter how useful doing them might be if judged solely from the vantage point of truth-seeking. Familiar instances include rules that the state cannot torture people to get confessions, nor can it require people to testify against themselves nor to submit to interrogation without an opportunity to have counsel present, although allowing such procedures might contribute to identification of malefactors. If the state were allowed untrammeled search and seizure of people's homes or cars, the police might catch more criminal perpetrators. If citizens were required to carry internal passports, fleeing felons might better be tracked. Such acts by the state are, however, proscribed. In part, these policies reflect doubt about the accuracy of coerced information and the excessive costs of massive surveillance. However, at their most basic level, they also express core value preferences: the a priori dignity and worth of the individual and the need to restrain government authority from violating that dignity. The exclusion from trials of evidence tainted by unacceptable government conduct expresses the strength with which these individual-protecting, government-restraining values are held.

Similarly, in consideration of the dignity and worth of the individual and the threat posed by unrestrained state authority, there are also some things the state *must* do. Thus, when it confronts the individual with the prospect of punishment, the state must do things to equalize the balance of power. It must afford a trial by a jury of one's peers. It must supply a state-paid attorney to accused persons who cannot afford their own. It must grant a right to confront accusers, access to compulsory process, and other such affirmative rights.

Finally, the adjudicative process is guided by preferences for some risks over others in the face of inevitable uncertainty. Thus, in criminal actions, the absolute value of the individual and the suspicion of unrestrained state power are encoded in the presumption of innocence to which all defendants are entitled. The well-known maxim—better ten guilty people go free than one innocent be mistakenly convicted—displays the intensity with which preferences for the dignity of the individual and restraint of state power dominate any desire purely to seek the truth. Analogously, in civil trials, a fundamental preference for an undisturbed status quo is effectuated by placing the burden of persuasion on plaintiffs with varying degrees of weight (preponderance of the evidence, clear and convincing proof) depending on what is at stake. Again, the rules reflect a system animated by the conviction that values other than truth matter.

Each of these value preferences is, of course, subject to debate and reconsideration. For example, concerns about escalating crime have caused some to conclude that current restraints on police and prosecutors and the latitude of defendants' protections are excessive. Proponents of such views argue that there is more to fear today from criminal perpetrators than from authoritarian government and that too exacting a protection of individual defendants may impose too great a cost to society.[5] Others maintain that recent reductions of defendants' rights

---

[5]*See, e.g.,* United States v. Leon, 468 U.S. 897 (1984) (holding that Fourth Amendment does not require exclusion of evidence obtained in reasonable reliance on a warrant ultimately found to be invalid). Reflecting the Supreme Court's recent tendency to increase police-prosecution powers at the expense of traditional defendants' rights, the *Leon* majority explicitly invoked the "social costs" of

undermine vital traditional values.⁶ Wherever the evolving balances on these issues are struck, no one argues that effects on truth-seeking are the only thing at stake in the legal system.

## Procedural Truth over Authoritative or Expert Truth

A second major characteristic of the American system is its considerable skepticism about authority-based or expertise-based claims to truth. Suspicion of established authority (monarchy, church, wealth, expertise) played a considerable role in the founding of the nation. Convictions about the risks of uncontrolled or unlimited power shaped our system from the outset. Power was structurally divided, its exercise subject to checks and balances. Practiced only in incomplete and uneven ways, the American commitment to individualism nonetheless informed institutions and norms. The frontier spirit glorified the practical generalist and the commoner-layperson at the expense of more "civilized" and expert elites. Commitments to cultural diversity and value pluralism, nascent and partial at best, nevertheless had some significance from the outset and intensified as the nation became more heterogeneous. Each of these strands is woven into the basic American belief that neither traditional nor expert-based claims to authority should be unreservedly embraced within the organized institutions of legal power.

Out of this matrix of ideas and traditions emerged a faith in process that transcends both particular fixed ideas about truth and claims that authority or expertise is the appropriate source of truth. One corollary is the conviction that at least in adjudication of disputes, truth is best sought and other important values best actualized through a process—the adversarial process. The adversary process expresses the conviction that just resolution (and therefore also functional, provisional truth) is best attained through a clash of viewpoints. Essentially, the system is entrepreneurial. Differing versions of disputed facts are presented before a neutral and passive decisionmaker who is required to make a determination about those facts. For the system to work, each contestant must have the incentive and the resources to find the facts and frame the arguments to support her version of truth. Any weaker party, such as the defendant in a criminal case, must have sufficient resources to perform this task. What appeared above as a demand to restrain government power and protect individual rights emerges again here as a necessary

having too many guilty defendants go free or receive reduced sentences. *Id.*, at 907.

⁶See, e.g., Justice Brennan's vigorous dissent in *Leon, id.* at 928. Justice Brennan argues that the majority's restriction of the exclusionary rule "ignores the fundamental constitutional importance of what is at stake here." *Id.*, at 929. He urges that "loss of that (excluded) evidence. . .is the 'price' our society pays for enjoying the freedom and privacy safeguarded by the Fourth Amendment. . .[C]ompliance with Fourth Amendment requirements makes it more difficult to catch criminals." *Id.*, at 941.

component of the law's procedural approach to truth/justice: The state must facilitate defendants' presentation of their side of the story.

Another corollary of process-oriented adversarial adjudication is the emphasis it places on lay judgment. Although legal and other experts have crucial roles in the resolution of disputes, laypersons retain a critical dispositive function. Judge-experts regulate the process, applying a body of rules designed to keep the competition fair. Lawyer-experts represent the contestants, shaping issues and arguments in light of prior decisions. Scientist-experts testify regarding facts they have observed and provide theoretical context and informed interpretation about what may be known or inferred on the basis of their specialized knowledge.[7] However, the parties themselves control the sources of knowledge about the disputed facts, and lay jurors determine the facts that ultimately decide the fate of individual litigants.

## DNA IDENTIFICATION IMPLICATES THE BASIC LEGAL NORMS

### DNA, Individual Rights, and Limited Government

Convictions about the centrality of the individual, the need to restrain state power, and the preference for some risks over others all have important implications for the controversy over the use of DNA identification evidence. When the risks of using DNA evidence are judged not simply in terms of scientific accuracy or truth value, but in terms of values about the legitimate use of government power to punish actual individuals, significant and special dangers accompany the uncritical adoption of DNA identification techniques. At a minimum, individuals deserve the maximum feasible protection against having liability imposed on the basis of inaccurate information. Furthermore, individuals have rights stemming from values the legal system holds equal or superior to truth. The impact of forensic DNA identification on these vital systemic values must be evaluated with care. Some of the issues have analogs in earlier debates about other scientific evidence; other problems are essentially unique to this procedure.

### *DNA Identification Techniques*

DNA identification is highly touted because each individual's DNA patterns are believed to be unique. The inference is that identification processes based on DNA are infallible. Although it may well be correct that each individual's genome is unique, that claim obscures certain realities about current DNA identification techniques, particularly as used in forensic settings.

---

[7]For a discussion of the role of the expert witness, see RONALD CARLSON ET AL., EVIDENCE IN THE NINETIES 519 (3d ed. 1991, The Michie Company, Charlottesville, Virginia).

Much of human DNA is the same in all individuals; indeed, much DNA is common to all species. However, only a small percentage of the total human genome is at present known and characterized. Studies tracking the prevalence of particular patterns within various subpopulations are sparse. Limited by these research realities, as well as the costliness of the analysis, DNA identification as currently carried out uses only a handful of probes to analyze a few genetic sites that are known to be characterized by significant individual variations. The very small amount of data actually generated by these probes is then read and interpreted for significance. Databases analyzed according to the principles of population genetics supply comparison data to enable what are essentially probabilistic statements regarding how likely it is that a particular declared match could be coincidental as opposed to being indicative of the presence or identity of a given individual. Thus, the assumed uniqueness of individual DNA patterns is not sufficient to validate present forensic identification techniques. Furthermore, all of the sample collection, analysis, and statistical interpretation that are part of DNA identification are designed and conducted by human beings. At multiple points and for diverse reasons, each of these processes is subject to error and debate regarding both performance and judgment, as are all forms of testing and laboratory procedure.[8]

*Errors in the Matching Procedure*

The potential for errors in the DNA identification procedure itself has been extensively documented in other parts of this book and elsewhere.[9] Numerous potential problems afflict the process (e.g., contamination or degeneration of the sample, difficulty or conflict in reading autorads). Many of these risks are severe. Any of them could produce an illegitimate result: legal liability imposed on the

---

[8] For data on laboratory error rates, see, e.g., EDWARD J. IMWINKELRIED, METHODS OF ATTACKING SCIENTIFIC EVIDENCE 3-7 (1982, The Michie Company, Charlottesville, Virginia) (reporting results of a study of forensic lab accuracy by the Law Enforcement Assistance Administration showing levels of error as high as 51% (paint analysis) and even 71.2% (blood analysis) in doing various forensic tests). More recent proficiency testing, including evaluation of the accuracy of DNA analysis, also presents a disturbing profile. *See* Randolph N. Jonakait, *Forensic Science: The Need for Regulation*, 4 HARV. J.L. & TECH. 109, 109-124 (1991) (summarizing forensic laboratory proficiency studies and concluding "a review of the data...indicates that laboratory performance is inadequate and unreliable.").

[9] *See, e.g.,* Hoeffel, *supra* note 3; William C. Thompson and Simon Ford, *DNA Typing: Acceptance of Weight of the New Genetic Identification Tests*, 73 VA. L. REV. 45 (1989) (assessing problems in techniques and statistics as used in adjudication); R.C. Lewontin and Daniel L. Hartl, *Population Genetic Problem in the Forensic Use of DNA Profiles*, 254 SCIENCE 1745 (1991) (arguing that population genetics data used to analyze DNA evidence are at present seriously deficient); Eric S. Lander, *DNA Fingerprinting on Trial*, 339 NATURE 501 (1989) (scientific analysis of DNA identification errors in several trials).

basis of inaccurate information. Although laboratory results can be retested and errors corrected, inaccuracies in forensic testimony could have determinative and irreversible effects on an individual's life.[10]

*Errors in Statistical Data Interpreting Significance*

Beyond analysis of the sample itself, the statistical judgments intrinsic to DNA identification are particularly vulnerable to error in ways that pose special problems for the legal process. At the outset, the very high numbers yielded by DNA matching evidence can be quite misleading. Thus, odds of 100 to 1 may sound like very persuasive evidence. Yet if there are many thousands or millions in whatever is the relevant subpopulation, 100 to 1 odds still leave a large number of individuals in the pool of potential defendants. Probably on the 100 to 1 level, defense attorneys can make that point and jurors can understand it. However, when the odds are jumped to the much more overwhelming levels occasioned by DNA identification, the same lesson may be harder to instill. The point is illustrated in a story told about a British case in which DNA evidence was used. Supposedly, the testimony alleged that on the basis of DNA evidence in sperm recovered from a murdered woman's body, the odds that the sperm came from defendant were about 6–8 million to one. (Put another way, only about 20 persons in England other than the defendant might have been the source of the sample analyzed.) Significantly, however, the probabilistic testimony against the defendant initially failed to discover and note that the woman's husband, as well as the defendant, was also one of the 20.[11,12]

In addition, errors in the calculation of the odds themselves may readily occur.

[10]Use of DNA identification techniques is too recent to have yielded, as yet, a set of errors documented to have played a role in convicting particular defendants. However, the potential for such errors is evident. A number of examples are discussed by Eric Lander, *supra* note 9, including *People v. Castro*, see *supra* note 3, in which Lander was an expert witness. In *Castro*, the prosecution offered testimony about very high odds of a DNA match between the victim's blood and blood on the defendant's watch. An unusually aggressive challenge to admission of this evidence eventually resulted in a joint statement by prosecution and defense experts that the offered evidence was unreliable. Consequently, it was not admitted. Replication of the extensive scientific and legal resources mobilized in *Castro* can certainly not be assumed, however. Without it, the flawed evidence would likely have been introduced and influential. Similarly, had the California Supreme Court not paid unusual attention to probabilistic evidence introduced in People v. Collins, 68 Cal. 2d 319, 438 P.2d 33, 66 Cal. Rptr. 497 (1968) grossly inaccurate statistical data would have played an important role in conviction of the defendants in that case. See *infra*, note 13. Practical logic suggests that errors will often escape intensive questioning and that errors not brought to light during the adversarial process itself are unlikely to be detected and corrected later.

[11]*See* Brian Sheard, *DNA Profiling*, 58 MEDICO-LEGAL J. 189, 197 (1990–91).

[12]*Editor's note:* This anecdote illustrates, among other things, aspects of Bayesian thinking emphasized in this and the chapter by D. Berry (this volume).

For instance, if relevant odds are multiplied on the assumption that variables are independent when they may not be,[13] the state may not be put to the test that the legal system demands. Errors resulting from statistical miscalculation may not be small; indeed, inaccurately assessed probabilities will likely be off by orders of magnitude. A prosecutorial or judicial guess about how to correct for such errors when they occur[14] seems deeply incompatible with the protection of individual dignity and restraint of state power that are incumbent upon legal proceedings.

The accuracy of probabilistic inferences regarding DNA evidence also depends on the appropriateness of the population samples used and of projections based on

---

[13]Although it did not involve DNA identification, problems in statistical evidence are illustrated in *People v. Collins, supra* note 10, (conviction reversed in part because statistical expert testimony used product rule without establishing independence of variables which were likely not all independent). For a critique of this case and of legal use of statistical evidence generally, see Lawrence Tribe, *Trial by Mathematics: Precision and Ritual in the Legal Process*, 84 HARV. L. REV. 1329 (1971). The product rule is also used in calculating probabilities of random match in DNA identification, usually without any empirical showing that the patterns of variation at the genetic sites probed are independent of one another.

[14]For example, in a recent New York case, the court, in the interests of due process, reduced the testified-to probability statistic for each test by a factor of ten to offset the possibility that the population used in the calculation was not in Hardy-Weinberg equilibrium. People v. Wesley, 533 N.Y.S.2d 643, 659 (1988). In a recent California case examining the application of the product rule to probability projections regarding DNA evidence, the appellate court stated that although the frequency figures challenged by defense experts might not be exact, their probative value was "very substantial" even if "a figure such as one in six billion is off by three or four billion." People v. Axell, 235 Cal. App. 3d 836,853, 1 Cal. Rptr. 2d 411,421 (1991). The remark suggests the court does not understand the multiplicative character of the product rule and the magnitude of errors that might result from its mistaken application. There is a "funny-money" quality to the numbers employed in DNA identification. For instance, in Spencer v. Commonwealth, 238 Va. 295, 384 S.E.2d 785 (1989), *cert. denied* 493 U.S. 1093 (1990), the court held there was no reversible error where stated odds of 135 million to 1 were increased to 705 million to 1 without written advance notice to the defense. In the Court's view, given that the defense asked for and got an overnight recess after receiving the change in numbers, the change represented no prejudice to the defense. The fact that a sixfold change in numbers was deemed not sufficiently significant to constitute prejudice illustrates the difficulty in applying ordinary concepts of meaningfulness to such large numbers. Finally, in State v. Nielsen, 467 N.W.2d 615, (Minn. 1991) the Minnesota Supreme Court held that even if DNA identification evidence was erroneously admitted, the error was "harmless" because the evidence did not have a significant impact on the verdict; the court deemed other evidence sufficient to produce the same result. In setting aside challenges to the appropriateness of the DNA evidence in this rape-murder scenario involving two brothers, the court stated in conclusory fashion that DNA from the crime scene matched defendant's but not his brother's. This statement seems to assume the accuracy of the very evidence that is being challenged. The court also does not discuss whether the testimony about the significance of the match (the odds of randomness) took account of the presence of the brother. Although the crime was appalling and the nonscientific evidence against defendant was strong, concerns about the reliability of the underlying procedure and of the accuracy of the stated odds (9.5 million to 1), as well as about the court's conclusion that the testimony played no significant role in affecting the conviction, are troubling at best.

them. If the population genetics data underlying testimony are such that probability numbers stated by expert witnesses and repeated by prosecutors are "unjustified and generally unreliable,"[15] then both the defendant's right to whatever protection truth offers and the entitlement to a presumption of innocence are severely compromised. The traditional preference, as noted earlier in this chapter, that it is better for ten guilty to go free than for one innocent to be convicted states the system's desire that the risks of error fall heavily in one direction rather than the other. Yet that ten to one ratio, sufficient to express a powerful point in ordinary numerical terms, makes the policy preference seem all but insignificant when viewed through the lens of probabilities articulated in terms like many millions to one.

## Errors in the Appropriate Uses of Statistical Evidence

Even if statistical judgments about the significance of a DNA sample match are appropriately calculated, resulting probability statements must be properly used. For example, it should be made clear to the jury that, at most, such statements allow an increased degree of confidence about the probable correctness of an already existing inference about guilt based on other evidence. This means that in the absence of significant other evidence, introduction of DNA evidence would be unfair. If, for example, there were no evidence against an accused other than a declared DNA match between material found at the scene of a crime and a sample located in a DNA data bank, use of the probabilistic statements about the significance of the DNA evidence would be inappropriate, even though the numbers alone would make such data seem enormously significant. Ultimately, this judgment reflects not only legal value standards but also the nature and limits of statistical probabilities. Statistics allow predictions about uncertainties; they do not establish facts.[16]

The potential problem for DNA-type evidence is illustrated by a case involving human leukocyte antigen (HLA) testing, a procedure less powerful than, but procedurally analogous to, that used in DNA identification. Because there has been more time for the legal system to assimilate evidence based on the HLA process than the DNA process, some comparisons between the two can be helpful. In *State v. Hartman*, a sexual assault case, the Wisconsin Supreme Court reversed the trial court's admission of testimony about the probability of defendant's paternity of complainant's child.[17] Although the court admitted testimony showing that the

---

[15]Lewontin and Hartl, *supra* note 9, at 1750.

[16]Much of the evidence used in legal proceedings is subject to the same observation. The point is not that such evidence should not be used, but that it should be used with care and an understanding of its limits. See discussion regarding court's treatment of cases involving only DNA or HLA evidence in Addendum *infra* at 52–54.

[17]145 Wis. 2d 1, 426 N.W.2d 320 (1988) (new trial ordered because statistical calculation of probability of paternity should not have been admitted; calculations inappropriately assumed that

defendant could not be excluded from paternity, it objected to testimony about the probability of paternity because those calculations began with what was described by the expert as a "neutral" assumption that there was a 50-50 chance that the defendant was the child's father. Consequently, the court reversed the conviction because the calculations that should not have been admitted assumed a fact, sexual access, that was itself the core element of the crime charged.[18] In essence, the testimony trampled the presumption of innocence.

In *Hartman*, there *was* evidence other than the probability of paternity calculations on which a strengthened inference of guilt could fairly have been based. But the expert (and the trial court) failed to clarify that the HLA-based evidence regarding probability of paternity could only legitimately be used to augment whatever prior probability the jury placed on defendant's guilt based on that other evidence. Reversal of the conviction tainted by the misunderstanding was therefore required.

The *Hartman* formulation makes visible the more fundamental problem of how to integrate mathematical evidence with a presumption of innocence. What should a presumption of innocence *mean* in mathematical terms? Should it mean, as the expert in this case seemed to assume by labeling his 50-50 odds as "neutral," that the chances are *equal* that the person is innocent or guilty? Or did the expert mean that the 50-50 odds reflected *his* assessment of the probabilities founded on other actual evidence in the case? Or that 50-50 odds simply represented one of a number of illustrative possibilities regarding prior assessments of guilt? Or should it mean, as yet others might argue, that merely being the subject of prosecution suggests that the defendant is likely guilty and that the only issue is whether the jury is persuaded of that guilt "beyond a reasonable doubt?"

On the contrary, the legal presumption is supposed to mean that a juror properly starts from an absolute assumption of innocence and may then be gradually moved by an accumulation of evidence toward a judgment of guilt. This is the presumption's meaning when the legal system selects jurors who must attest to their lack of bias. Yet given systemic requirements such as probable cause for investigating and indicting a suspect, an assumed probability of zero regarding guilt is arguably neither realistic nor appropriate. Here lies the nub of the problem; there is intractable conflict between fundamental social values as embodied in legal rituals and precise quantifications about probable truth. Partly for these reasons,

defendant committed the crime). A closely analogous case was reversed on the same grounds by a New Jersey appellate court. State v. Spann, 236 N.J. Super. 13, 563 A.2d 1145 (1989), *cert. granted*, 122 N.J. 376, 585 A.2d 381 (1990).

[18]On the facts of the case, consent was irrelevant. The defendant was accused of sexual assault of a 14-year-old minor who could not legally consent. Similarly, in *State v. Spann, supra* note 17, defendant was a corrections officer, for whom sexual relations with the complainant, a prisoner, were illegal whether or not she consented. The *Spann* court explicitly mentioned that the testimony violated the presumption of innocence.

some have argued that any mathematical quantification of legal standards like the presumption of innocence or the standard of proof is unacceptably dangerous to the values underlying the legal system.[19]

Finally, adjudicative decisionmakers must understand the difference between confirming or refuting a DNA identification and determining guilt or innocence. Even if there is a very high probability that the DNA in question came from a given person, that does not mean that the person is guilty of the crime charged. There may be other reasons for presence of the person, or for that matter, of the DNA. A sample (hair, fingernail, drop of blood) might be planted, like a gun. The person might have been present and might therefore be the source of a fingernail, hair, or semen, but might still not be the one who committed the crime. (An individual might have been trying to rescue a crime victim when hair was shed or blood was spilled, for instance).[20] These latter possibilities, of course, vary depending on the factual circumstances, the location from which the DNA sample was taken, and the crime charged. Nevertheless, there is serious danger if such possibilities are forgotten in the rush of numbers. Particularly worrisome is the possibility that decisionmakers will mistakenly conclude that figures about the odds of a random match can be uncritically transformed into a finding about guilt or innocence on the theory that such large numbers almost automatically seem to satisfy the required finding of proof beyond a reasonable doubt.

*Access to Samples from Suspects*

Legal values about individual dignity are also encoded in policies that restrict what the government can do to gather evidence or identify suspects. Because DNA identification depends on matching genetic material from a crime scene or victim with material from an alleged perpetrator, there are important questions regarding the government's right to access such material from a suspected individual. Courts will initially analyze such issues under fairly well established Fourth Amendment guidelines governing search and seizure.

At the outset, differences in factual characteristics should yield differing

---

[19]*See generally* Tribe, *supra* note 13. (With a few possible exceptions, "the costs of attempting to integrate mathematics into the fact-finding process of a legal trial outweigh the benefits." *Id.* at 1377.)

[20]Such concerns cut both ways, of course. For example, despite testimony for the defense by an FBI crime technician that DNA in semen on a rape victim's clothing did not match that of the defendant, the jury convicted. See Jack Ewing, *Conn. Jury Disregards DNA Test*, NATL L.J., April 23, 1990, at 9. The scene of the crime was a notorious lover's lane, and multiple sources of semen were present at the location. Proof that one such sample tested did not match the defendant did not, in the minds of the jurors, and for that matter in the mind of another FBI specialist in DNA identification, establish that he did not rape the victim. (Personal conversation with Dr. Bruce Budowle, FBI, Feb. 1991).

judgments about the legal permissibility of even certain behaviors that may at first glance appear analogous. For example, exigency regarding preservation of evidence has traditionally justified some kinds of search of body or body products without a warrant or with reduced criteria of probable cause. Such justification would normally be absent in the instance of DNA samples because, unlike blood alcohol levels, for instance, DNA will not change or disappear. Consequently, there should be no relaxation of standards of probable cause before a genetic sample may be demanded from a suspect.

More basic issues are also implicated. Much of the doctrine governing search and seizure has focused on the degree of physical intrusiveness as a variable limiting governmental access to a person's body. Because a person's DNA can be accessed from so many different sources, many of them minor or comparatively external to the body, the issues of privacy and dignity presented by such access may appear comparatively trivial—analogous to established and accepted techniques involving, for example, tests of alcohol in breath or blood or the taking of fingerprints. Because they are in some sense in public view, for instance, fingerprints carry no protectable expectation of privacy. Access to DNA in hair or fingernails might be thought to be similarly unobjectionable. Moreover, the ease of access differentiates DNA from more extreme intrusions that courts have barred, such as surgery to remove a bullet from the body or forced regurgitation of the contents of the stomach.[21] Such comparisons would be misleading, however. Because of the unparalleled breadth and extraordinary salience of data that can be derived from DNA (including such factors as the individual's parental and genealogical history, present and future health status, race, and sex), literal physical intrusiveness may not be an adequate yardstick regarding the invasiveness of privacy and dignity that result from DNA identification techniques. More restrictive tests governing unconsented recovery of DNA material from a suspect may be needed.

## *Access to Samples from the General Population*

In addition to the issues raised by collection of samples from particular suspects, significant problems are presented by general data banking of DNA information. Such data banks might be compiled to facilitate searches for particular matching samples, for purposes of research on the distribution of genetic patterns in the population, or for purposes unrelated to the criminal law.[22] With regard to such data

---

[21]*See, e.g.*, Winston v. Lee, 470 U.S. 753 (1985) (unconsented surgery to remove bullet for criminal evidence not constitutionally permissible); Rochin v. California, 342 U.S. 165 (1952) (forced use of emetic to empty stomach not constitutionally permissible).

[22]For instance, the Defense Department recently announced its intention to establish a data bank of DNA samples from every member of the military service to aid in identifying the remains of war dead. Warren E. Leary, *Genetic Record to Be Kept on Members of Military*, N.Y. TIMES. Jan. 12, 1992, at 15.

banks, broad legal policy goals pull in conflicting directions, creating an ironic paradox. A major concern about the use of DNA identification in particular trials is that the data banks used to calculate the statistical significance of a match are often too small and too undifferentiated to yield reliable data. In particular, a failure to use databases appropriate to various subpopulations can yield outcomes that are seriously inaccurate. Moreover, population genetics organizes data on the basis of racial and ethnic subgroupings. If inaccuracies are more severe in some subgroups than in others, actions based on them may constitute de facto race or national origin discrimination as well. Prevention of such discrimination is a central legal norm. Consequently, data banks must be assembled and used in ways that vindicate this important legal concern. Thus, concerns about both accuracy and conformity to antidiscrimination principles support the expansion of DNA data banks.

Yet the very process of deepening and extending data that could create fairer and more accurate statistical data for use in individual trials also increases the problem of invasion of privacy and other potentials for discrimination against racial, ethnic, or genetically "different" categories within the population generally. Subjecting more people to the sampling procedure dilutes its potential for stigma, thereby reducing some concerns about invasion of privacy. However, simply being subject to the procedure is not the only issue. Rather, it is the sweeping potential for use and abuse of the data acquired that raises the most serious problems. Given the potential uses of DNA information by sources as varied as crime control agencies, life and health insurers, and employers, widespread stockpiling and uncontrolled use of such information seems risky. Some argue that if everyone's DNA is analyzed, notions of disability or impairment will be neutralized because everyone will have some "abnormal" genes. However true this may be as a matter of science and description, socially constructed notions of good and bad, of normal and abnormal, will still engender selective stigmatization and priority setting on the basis of genetic endowment. Such outcomes are deeply problematic for the individual-protecting, government-restraining values of the legal system.

Privacy and accuracy concerns regarding data banking seem doomed to be in tension with one another. Analogous conflicts sometimes lead the legal system to conclude that societal interests in truth must give way to values such as individual privacy. However, even though the state's interest in accurate information for prosecution may be overridden by competing values, the individual's interest in avoiding punishment that stems from inaccurate information remains extremely strong. If concerns about invasion of privacy keep databases small enough that projections regarding significance are comparatively unreliable, the risk that errors could harm individuals might become unacceptably great. Thus, the asymmetrical weighting of truth-value according to function might yield a policy of refusing any use of DNA evidence, rather than allowing use that may unacceptably put individuals at risk.

## DNA and Procedural Rather Than Substantive Truth

The second major characteristic of the legal system is its commitment to procedural rather than substantive forms of truth. Dovetailing with the individual-protecting and government-restraining values is an equally fundamental suspicion of authoritative claims to truth. The American political tradition endorses value pluralism and rejects absolute authority, whether secular or religious. In the absence of sources that could authoritatively establish substantive truth, the legal system eschews expert truth-finding and adopts procedural approaches instead.[23] Free speech and democratic debate are procedural means of pursuing truth in the political sector. Analogously, for purposes of resolving disputes and establishing norms of conduct, the legal system employs an adversary clash of views in a process that assigns important roles to laypersons as well as to professionals and experts. The legal system's procedural approach to truth, its skepticism about absolute authority, and its trust in lay judgment have important implications for the debate about use of DNA identification techniques in court.

### *Adversary Resource Balancing*

For the adversary process to function properly, each side much have adequate resources to present its best possible case. In particular, where the government accuses an individual of a crime, if the individual lacks resources to balance the disparity of power, the attainment of legitimate outcomes through the competitive process will be compromised.[24] This resource-balancing principle raises special issues in the context of DNA identification.

In early cases presenting DNA evidence, courts tended to admit the evidence readily. Prosecutors, basking in the aura of prestigious science, presented testimony about overwhelming quantitative evidence that seemed to settle cases beyond any reasonable doubt. Experts were readily available from biotechnology firms eager to validate their proprietary techniques in this important new market. The defense bar was unprepared to challenge the new techniques. Defense experts willing to question "the general scientific acceptance" of methods that rested not merely on forensic needs but on dramatic breakthroughs in basic science were in short supply.

---

[23]In contrast, the European inquisitorial system assigns a much more central role to the judge as a professional truth finder. For comparison of the two systems, see, e.g., STEPHEN LANDSMAN, THE ADVERSARY SYSTEM DEFINED 48–51 (1984, University Press of America, Lanham, Maryland).

[24]From this root, for example, grows the constitutional requirement that government must provide a lawyer if the defendant cannot afford one.

If ever there was a time for concern about balancing resources, surely this was it. Yet, when offered the tantalizing certainty DNA appears to provide, some courts forgot basic adversary principles. Rather than worrying that one-sidedness was depriving them of the competitive analysis essential to procedural truth-finding, they apparently concluded that an absence of defense experts established the technique's reliability.[25] Although more literature and personnel analyzing flaws of DNA identification have since appeared, the degree of imbalance between prosecution and defense expertise and resources remains a serious problem.

The resource-balancing principle also mandates that the defense has a right to fully prepare its own case and to cross-examine the prosecution. As a by-product, the defense has a right to be notified of and to inspect prosecution evidence. Because of the large start-up and analysis costs, as well as the potential for profit, commercial firms conducting DNA tests for the prosecution have sometimes sought to bar or limit the scope of defense access by claiming trade secret or other proprietary protection for their procedures and protocols.[26] Particularly where commercially based experts have played critical roles in validating DNA evidence, such claims should generally be denied. Even the traditional compromise of granting protective orders that prohibit disclosure to individuals other than defense personnel undermines the ability of scientists in general to evaluate particular techniques and protocols of this new methodology.

The defense generally also has a right to access any physical evidence that the prosecution uses. Yet a forensic DNA sample might be so minute or badly tainted that the usable sample is entirely consumed by prosecution analysis, leaving none for defense cross-checking. Given the pivotal nature of DNA evidence, perhaps standards evaluating the admissibility of evidence where the sample was unavailable for defense examination should be more stringent than standards for other types of evidence.[27] Another consequence of the resource-balancing principle

---

[25]*See, e.g.,* cases discussed in Stephen C. Petrovich, *DNA Typing: A Rush to Judgment,* 24 GA. L. REV. 669, 671 n.14 (1990); Anthony Pearsall, *DNA Printing: The Unexamined "Witness" in Criminal Trials,* 77 CAL. L. REV. 665, 691 (1989); Hoeffel, *supra* note 3, at 499 n.193 and 511. *Cf.* Paul C. Gianelli, *The Admissibility of Novel Scientific Evidence: Frye v. United States, a Half-Century Later,* 80 COL. L. REV. 1197, 1243 (1980) (describing courts' admission of voiceprint evidence after inferring that an absence of defense experts meant a technique was generally accepted by the scientific community).

[26]*See, e.g.,* State v. Schwartz, 447 N.W.2d 422, 427 (1989) (mandating broad requirements for disclosure to the defense; failure by Cellmark to supply requested information in this case violated standards). *See also,* Hoeffel, *supra* note 3, at 502. According to newspaper reports, Timothy Spencer was denied access to Lifecodes' laboratory notes and other internal documents during his trial in which he was sentenced to death with the help of DNA evidence. *See* Alan Cooper, *DNA Case Is First Before a State High Court,* NAT'L L.J., July 3, 1989, at 14 (discussing the *Spencer* cases, 238 Va. 275, 384 S.E.2d 775; 238 Va. 295, 384 S.E.2d 785; 238 Va. 563, 385 S.E.2d 850 [1989]).

[27]Most courts have held there is no constitutional violation if the consumption of the sample by the test was necessary. In such instances, however, they may require notice to allow a defense expert to be

is the state's obligation to pay for defense procedures and experts. That right has often been narrowly construed in practice. Given the high costs associated with DNA evidence and the unusual salience of the evidence if used, should that obligation be greater than ordinary in this instance? Indeed, should the state perhaps be required to pay for defense analysis of DNA evidence for exculpatory purposes as well as for challenges to inculpatory evidence?

An even more difficult question arises at this juncture. Stated at its most general level, the resource-balancing policy mandates a wide and comparatively unfettered right for defendants to present their cases at the same time that limits are placed on prosecutors to guarantee the protection of individuals. For example, the prosecution must notify the defense of and provide access to prosecution evidence. It must also turn over any potentially exculpatory evidence it discovers. No reciprocal obligation is placed on the defense.[28] Does this suggest that the overall standards for defense admission of potentially exculpatory DNA evidence ought to be less arduous than standards for its use by the prosecution? Standards for admissibility of evidence are traditionally unitary. More will be said about the specifics of those standards shortly. The question is, whatever the standard adopted, should that standard vary to reflect the resource-balancing principles of the adversary system? One commentator has suggested that requirements regarding the adequacy of laboratory protocols should be higher for the prosecution, given its higher burden of proof, than for the defense, which needs only to establish reasonable doubt.[29] At least, standards for prosecution use ought not to be lower than for defense use.[30] DNA evidence pointing to exclusion of a particular defendant is generally more scientifically reliable than is evidence of inclusion or identification. Thus, the argument that standards for DNA admissibility should vary consistently with legal value preferences is made even stronger by the scientific characteristics of this particular technique.

present. *See* discussion in Addendum, *infra* at 49; Hoeffel, *supra* note 3, at 524. The problem of too small a sample for adequate checking by both sides will likely be less severe as PCR (polymerase chain reaction) techniques that allow extensive copying of DNA material become more accessible.

[28]*But see* California's recent adoption through the popular initiative process of unusual "mandatory reciprocal discovery" that places just such a burden on the defense. California Proposition 115, *adopted* June 5, 1990, *codified as* CAL. CONST. ART. I, Section 30(c).

[29]*See, e.g.,* Gianelli, *supra* note 25 at 1248 (proposing that due process rights and right to compulsory process should include right to present evidence by a preponderance-of-the-evidence standard for defense but a beyond-a-reasonable-doubt standard for prosecution). To my knowledge, no court has adopted Gianelli's proposal and one has rejected it. U.S. v. Jakobetz, 955 F.2d 786 (2d cir. 1992). *Jakobetz* is discussed in the Addendum to this chapter, *infra* at 45.

[30]*See* Eric Lander's description of a New York prosecutor's objection to defense use of DNA evidence as "inaccurate, and therefore, unreliable" in a trial held a mere three months before the same DA's office sought to introduce DNA evidence in the *Castro* case. Lander, *supra* note 9, at 505 (1989).

This array of issues must be adequately addressed and resolved in the special circumstances presented by DNA evidence. Because DNA identification evidence is likely to be particularly decisive where it is used, policies about resource balancing may need to be differently weighed in this context than in traditional evidentiary settings.

*Lay Influence in Adversary Decision Making*

The adversary system's institutionalized skepticism about authoritative or expert determinations of truth is also reflected in its assignment of critical discretionary judgments to lay decisionmakers. Legal experts—judges and lawyers—have significant roles in adjudication. Judges apply the rules that structure the adversary process. Lawyers present and argue alternative versions of the law and facts, seeking to persuade and educate the relevant decisionmakers. Nonlegal experts, including scientists, testify where necessary to aid in the presentation and evaluation of relevant evidence. However, the constitutional guarantee of a right to a jury of one's peers ensures that lay jurors retain final and largely nonreviewable[31] control over particular outcomes. DNA evidence raises special problems for the balance of expert and lay decision making in the legal system.

*Tests of admissibility.* Legal rules governing the admissibility of evidence are decided by statute or precedent for each jurisdiction and are applied by judges to particular problems and cases. Once evidence is admitted by the judge, its probative weight, along with the credibility of witnesses, is a matter within the province of lay jurors. If evidence is not admitted, the jury will never hear it.[32] Thus, admissibility decisions, and the standards under which they are made, affect the jury's access to scientific testimony as well as the distribution of decisionmaking between judge and jury.

When a new form of evidence is first offered and challenged in a given state, that jurisdiction will decide how to determine its admissibility. The traditional test for admissibility of scientific evidence has been the *Frye* test, named for the case in which the standard requiring general scientific acceptance of a technique was established.[33] Like most legal rules, the *Frye* test is not self-executing. At a minimum, it requires determinations regarding *whose* acceptance and *of what* is necessary in order to satisfy the requirements. As applied to DNA evidence, for example, if a *Frye* inquiry asks whether research scientists accept the basic theory of DNA then

---

[31]Nonreviewable except when a judge grants a motion for judgment notwithstanding the verdict of the jury, something that happens only rarely.

[32]Unless, of course, the judge's decision is reversed on appeal.

[33]*See* note 1, *supra*. For more detailed discussion of the resolution of admissibility issues by various courts considering DNA evidence, *see* Addendum, *infra* at 39–47.

all such evidence would be admissible. If, on the other hand, scientific acceptance is deemed to require validation of a particular protocol (including what probe is used, what criteria are employed to declare a match, what databases are available, what statistical methods are adopted to project the significance of a match), a *Frye* inquiry may be extensive and its outcome far more problematic. Although courts in different jurisdictions have varied in their approaches and stringency, most using the *Frye* test have adopted a level of analytic abstraction between these two extremes.

In addition, some courts and jurisdictions have moved away from the *Frye* test, instead assessing scientific evidence under the comparatively loose standard of relevance or under a somewhat more stringent balancing standard of "helpfulness"[34] that weighs the scientific reliability of offered data against the possibility of their misleading or overwhelming the jury. This latter standard is both looser and tighter than the *Frye* test in that it dispenses with questions about scientific or forensic acceptance, but factors in the potential for jurors to overvalue certain types of evidence.

Whatever standard they employed, the earliest courts to confront DNA evidence generally admitted it with little question.[35] More recently, however, some courts have heightened and expanded the admissibility inquiry regarding DNA identification, and some have refused to allow offered evidence to go to the jury.[36]

The scope and stringency of admissibility tests will influence the quantity, type, and balance of lay influence, as compared to expert/professional (both scientific and legal) influence, in cases using DNA evidence. Refusals to admit offered evidence reflect judicial judgments that reliability is too low or prejudicial effect is too high. If DNA admissibility decisions apply more stringent standards than are applied to other types of evidence, traditional legal values are implicated. Comparatively greater refusal to admit DNA evidence will result in lesser influence for scientific expertise and in greater judicial influence, as compared to jury influence, over outcomes. The central issue is whether such shifts are desirable.

*Significance of shifts in expert and lay roles.* The evaluation of shifts in lay and expert roles turns in part on who wants to use the information that might be excluded. For example, exclusions of DNA evidence resulting in a reduction of lay/juror influence as compared to judge/professional influence may be particularly problematic where the defense seeks to use the evidence for exculpatory purposes. The situation is somewhat reminiscent of judges' retention of judicial control over

---

[34]This standard, based on Federal Rule of Evidence 702, was articulated in United States v. Downing, 753 F.2d 1224 (3d Cir. 1985). The case involved psychological testimony regarding eyewitness evidence, but the standard has been applied in some DNA cases as well.

[35]*See, e.g.,* Andrews v. State, 533 So. 2d 841 (Fla. App. 1988); Spencer v. Commonwealth, *supra* note 14.

[36]*See* cases cited, *supra* note 3.

the interpretation of written contracts. Analytically, such matters would seem to be questions of fact that would normally be entrusted to jury deliberation. However, a combination of jurors' actual historical illiteracy and the suspected bias of uneducated lower classes against educated elites caused judges in earlier eras to reserve greater power regarding written documents than would ordinarily seem warranted. Analogously, if defense DNA evidence were excluded on the grounds that jurors were "scientifically illiterate," the resulting diminution of jury role could be viewed as unfairly reducing the peer input constitutionally guaranteed to defendants.

More typically at present, exclusion of DNA evidence will deny the prosecution the opportunity to get scientific testimony before the jury. Those who are concerned about oversolicitous protection of defendants and increasing crime may well be disturbed if judges prevent inculpatory scientific expertise from reaching lay juries who will ultimately determine outcomes.

The much more critical issue, however, is who can best use and evaluate DNA evidence. A decision to heighten the judge's role in DNA cases expresses concern that jurors may not be able adequately to evaluate the weight of such evidence and experts. Certainly, jurors may have all the problems of evaluating DNA evidence that were identified above. They may fail to understand or appreciate the risks of error in sample collection, laboratory procedures, population genetics data, or statistical techniques. In particular, they may not understand that an error in one step of the process can cause final numerical estimates to be off, not simply by a little, but by orders of magnitude.[37] They may not understand how to integrate legal values with mathematical data. In particular, they may miss the important difference between the probabilities that a match is significant rather than random and the different issue of how those probabilities bear on the ultimate issue of guilt or innocence. Within the legal context, such errors bear not on an inaccurate theory or a faulty footnote, but on an individual's liberty and future. The potential for accurate inculpation and exculpation is magnetic; the risks, however, are great.

Jury use of DNA identification evidence may in some sense be trouble squared. First, DNA identification is based on complex and sophisticated techniques and employs very difficult statistical projections. Its validity is therefore extremely difficult to assess. Second, it presents seemingly "hard" quantified data in fairly simple endpoint terms, generally offering odds that are the equivalent of a numerical sledgehammer. The risk is that jurors will be enormously influenced by evidence they think they understand, when in fact they are unable to assess its accuracy and significance. This troubling combination may well be sufficiently worrisome to demand that more stringent tests for the admissibility of DNA

---

[37]A prominent scientist involved in several major cases on behalf of the defense estimated that a potential for at least 8000-fold error existed in calculation of the odds of match under the methods used by Lifecodes, one of the major companies presenting DNA identification evidence in court. *See* Lander, *supra* note 9, at 504.

evidence be applied. Perhaps it may be necessary to bar such evidence entirely, at least at present levels of accuracy and capacity for interpretive understanding.[38]

Several important caveats should be noted, however, before concluding that lay jurors should not hear DNA identification testimony, or should have their access comparatively restricted. First, there is evidence that jurors are not as overwhelmed by scientific testimony as some might think.[39] Moreover, decisions to prevent information from going to juries are typically made by judges. Where science is concerned, judges, too, are laypeople, not necessarily more skilled than jurors at evaluating the reliability of DNA evidence. Moreover, even scientific experts disagree about the evaluation of scientific and statistical data. Adversary presentation and cross-examination before the jury is the traditionally preferred legal remedy for controversies regarding truth.[40] However, despite their lay status with reference to scientific and mathematical data, judges *are* legal professionals experienced in assessing legal concepts like relevance and prejudice. They have responsibility for protecting individuals' rights against the risks of jury error or prejudice. Moreover, lawyer skill at presenting or questioning expert testimony will be the single most critical factor shaping its influence on either judge or jury decisionmaking. Thus, more than allocation of judge-jury function, the equalization of defense and prosecution lawyering competencies is the critical variable in fair and effective use of DNA evidence.

Second, ability to evaluate DNA identification data is not an issue occurring within a vacuum. A given type of evidence must be compared with alternative types of evidence that will be used by a decisionmaker in its stead. Other types of evidence, too, are subject to human error. Data about forensic laboratory accuracy regarding other forms of evidence suggest startlingly high levels of error.[41] Eyewitness testimony is often inaccurate, yet is typically accorded great weight by

---

[38]*See generally* Tribe, *supra* note 13 (analyzing in a pre-DNA context the complexities of statistical proof in adjudicative contexts and recommending such data generally not be used). Most commentators call not for barring DNA evidence but for withholding its use pending development of more thoroughly standardized and validated procedures.

[39]*See* Edward J. Imwinkelried, *The Standard for Admitting Scientific Evidence: A Critique from the Perspective of Juror Psychology*, 28 VILL. L. REV. 554 (1982–83) (arguing that available evidence calls into question the assumption that scientific testimony overwhelms lay jurors). For an interesting example of a jury not being swayed, see Ewing, *supra* note 20 (jury convicts despite FBI testimony for the defense of non-match).

[40]Note, however, that experts may be confused and overwhelmed by the adversary process. Not used to having constraints on their ability to say what they believe to be true, experts confronted with questions, objections, and evidentiary rulings governing the scope of their testimony may find it difficult to testify in a way they believe is accurate. (I am indebted for this insight to Professor Dan Rubinfeld who often testifies as an expert witness.).

[41]*See* sources cited in note 8, *supra*.

juries.⁴² A major task of lay jurors is to judge credibility of conflicting witnesses, yet recent empirical studies demonstrate that although people think they can tell when someone is lying, most people are no more accurate in judging veracity than if determinations were made by flipping a coin.⁴³ Juries base important assessments about a case on inference and on grass roots assessments of probability. The fact that those judgments are not formally quantified does not remove them from the domain of probabilistic assessment, and many of the same problems affecting the use of DNA identification afflict alternative sources of persuasion as well.

These points suggest that DNA evidence as assessed by jurors may be no worse and perhaps better than alternate forms of evidentiary decisionmaking. However, there is yet another layer of offsetting considerations. For example, both judges who hear and lawyers who present and cross-examine scientific testimony of varying types are repeat players in the legal system. With time and exposure, both will undoubtedly get better at using and critiquing DNA identification evidence. In contrast, lay jurors have dramatically fewer opportunities for increasing their sophistication about these techniques.

Moreover, with regard to jurors' use of alternative types of evidence, lay people may be more aware and skeptical about the shortcomings of other techniques such as voiceprint, lie detectors, or psychiatrists. As a result, they may give such testimony considerably and appropriately less deference than they might accord to DNA-type techniques that possess more of the aura of scientific infallibility. Similarly, jurors may also be at least somewhat aware of the fallibility of their own garden-variety inferences about evidence and credibility. They may therefore, to some degree, discount those inferences accordingly.⁴⁴

Finally, in developing commonsense inferences, laypeople are far less likely to think in the extremely large numbers projected by statistical testimony about DNA identification. Accordingly, even if erroneous, such "seat-of-the-pants" probabilistic assessments may be less damaging than an incorrect expert judgment stated in terms of odds of hundreds of millions. Indeed, erroneous homegrown probabilistic assessments may be less likely to produce a wrongly derived judgment than are even accurate data provided by experts if the latter data are incorrectly understood

⁴²*See* ELIZABETH F. LOFTUS, EYEWITNESS TESTIMONY (1979 Harvard University Press, Cambridge, Massachusetts) (summarizing theoretical and empirical literature on problems in eyewitness testimony).

⁴³*See* Daniel Goleman, *Non-Verbal Cues are Easy to Misinterpret*, N.Y. TIMES, Sept. 17, 1991, at C1 ("most people believe they can usually catch a liar in the act" but "there was no relationship between how much confidence people expressed in their ability to detect lies and their actual performance.").

⁴⁴*But see* material cited in note 43, *supra*, and sources cited in Imwinkelried, *supra* note 39, at 570 (summarizing some data regarding jurors' tendencies to credit lay testimony more than they should, particularly in comparison to scientific evidence).

or used, as, for example, where overwhelming probabilities of a DNA match are incorrectly transmuted into probabilities of guilt.

Although these considerations do not all point in a single direction, they do suggest that greater than ordinary caution in admitting DNA testimony for jury evaluation may be warranted. Despite the possibility of some shift in judge-jury function, and the possibility that scientific data will be denied to the jury, the risks of prejudicing the fate of an individual defendant are major indeed. Such risks are too great until the system has learned to cope more skillfully with the new techniques. Careful attention should be given to education of judges and lawyers, perhaps by independent scientific and statistical consultants. Increased understanding of the techniques could guide judges' decisions and could enable lawyers to educate both judge and jury more effectively.

## CONCLUSION

Despite the enormous potential for DNA identification to play an important role in adjudication, prudence dictates more than ordinary care regarding use of these techniques. The scientific community, independent of prosecutional or proprietary interests, needs to develop and validate standards for all aspects of the forensic procedure from sample collection to laboratory analysis to statistical projection. Without review and standardization, use of DNA identification techniques in legal adjudication is dangerous. Moreover, even with improved accuracy, there is reason for caution in legal adoption of DNA evidence. The issue is not one of opposition or conscious ignorance regarding scientific knowledge. Rather, it is one of differences in function and value as between different communities.

The legal system is not designed as an abstract search for truth; rather, it decides consequential outcomes for real individuals. In light of its role and norms, the legal system places significant weight on values other than truth, values that embody notions of justice as well. Use of DNA evidence must be shaped in ways that are compatible with individual dignity and limited government power. The legal system also embraces an adversary process that weighs competing versions of truth rather than deferring to authoritative or expert declarations of truth. Only when both parties in a legal controversy have adequate resources to contest the value of DNA evidence can concerns about prejudice and error be reduced and the roles of expert and lay decisionmakers be appropriately balanced. Meanwhile, the traditional skepticism about authoritative and expert claims to truth combines with the individual-protecting, government-restraining policies central to the legal system to legitimate a thoroughgoing wariness about uncritical embrace of DNA evidence.

## ACKNOWLEDGMENTS

I am grateful to Eleanor Swift and Dan Rubinfeld for their helpful comments on an earlier draft of this paper. I am also indebted to David Gass, Michael Ostrove, Tay Via, and Marie Villafana for their valuable research assistance.

# Addendum
## An Analysis of Decisional Law Governing the Use of DNA Evidence (As of January 1992)

**DAVID A. GASS[1,2] AND MARJORIE MAGUIRE SHULTZ[2]**

[1]Marshall, O'Toole, Gerstein, Murray, and Bicknell
Chicago, Illinois 60603
[2]Boalt Hall School of Law
University of California, Berkeley
Berkeley, California 94720

### Different Standards for Admitting DNA Fingerprinting Evidence in U.S. Courts

In the United States, a judge initially decides whether to admit or exclude DNA evidence from a trial, thereby determining whether a jury will hear such evidence. Legal standards used in making such a ruling are appealable for review by higher courts. Standards for admissibility affect crucial system values, including the balancing of truth-finding with other norms, the fairness of the adversary process, and the allocation of lay and expert roles. Different states and federal courts have used a variety of approaches to decide when DNA evidence meets a threshold level of reliability such that a jury should be allowed to consider it. We review here some of those approaches.

### Early Cases

Some courts first admitted DNA evidence in criminal trials largely because no experts testified against it. For example, in *Cobey v. State*,[1] a Maryland appellate court affirmed a trial judge's admission of DNA evidence in a rape trial because the evidence was generally acceptable in the scientific community. The appellate court deemed it "significant" that Cobey, the defendant, had not produced any expert testimony challenging the validity of restriction-fragment-length polymorphism (RFLP) technology. Cobey had averred that the database used by Cellmark was insufficient to support the statistical conclusions drawn from it, but again the court noted that no expert testimony supported this contention.[2]

---

[1]80 Md. App. 31, 559 A.2d 391 (1989).

[2]*See also* Andrews v. State, 533 So.2d 841, (Fla. Dist. Ct. App. 1988). A Florida appellate court affirmed a rape conviction, rejecting appellant's challenge to the admission of DNA matching evidence and probability evidence (839.9 million to 1). The state had several qualified experts testify, whereas the defense had none and relied solely on cross examination of the state's experts to discredit the DNA

## The Three-prong Frye Test as a Barrier to the Admissibility of DNA Evidence

Jurisdictions that have scrutinized DNA evidence most carefully before allowing its presentation to the jury usually have applied a "three-prong" *Frye* test.[3] *People v. Castro*[4] is a leading example, even though it comes from a lower court. The *Castro* court, applying a three-prong *Frye* test, was the first to reject DNA matching evidence in a criminal case. During a "*Frye* hearing," held outside the presence of the jury, each side may present scientific expert testimony to the judge. The *Castro* hearing lasted 12 weeks and generated 5000 pages of transcript.

Prong one of the *Frye* test asks, "Is there a theory, which is generally accepted in the scientific community, which supports the conclusion that DNA forensic testing can produce reliable results?"[5] The overwhelming trend among all courts (including *Castro*) has been to answer in the affirmative.

Prong two of the *Frye* test next asks, "Are there techniques or experiments that currently exist that are capable of producing reliable results in DNA identification and which are generally accepted in the scientific community?"[6] In some early cases, opponents of DNA evidence contended that generally accepted research techniques might not transfer well to forensic work.[7] However, prong two has also

evidence. The court saw two indicia of reliability that critics today might question. First, it noted that the DNA technology had been extensively used in nonjudicial settings for 10 years. It did not mention that forensic applications differed from conventional laboratory DNA analysis—a contention of critics of DNA fingerprinting. Second, it was impressed by the testimony of one of the state's experts that in reading a great many scientific journal articles and publications with regard to DNA testing, he was unaware of any that argued against the test's reliability.

Even with a pretrial admissibility hearing, a defendant may not be adequately equipped to challenge the state's DNA evidence. *See* Smith v. Deppish, 807 P.2d 144 (Kan. 1991). The Kansas Supreme Court, in its first examination of DNA fingerprinting in a criminal appeal, noted that the defendant at the *Frye* hearing did not use an expert to challenge the reliability of the specific test performed; did not challenge the qualifications of the State's expert witnesses; did not refute the expert testimony of the state; did not offer evidence that the DNA tests were unreliable; and did not offer evidence that the tests were not generally accepted by the scientific community. The Kansas Supreme Court found no error in the trial court's admission of the DNA profiling evidence under *Frye*. If the defendant had been armed with qualified experts of his own, the result might have been different.

[3] Frye v. United States, 293 F. 1013 (D.C. Cir. 1923) (establishing standards for admissibility of novel scientific evidence).

[4] 545 N.Y.S.2d 985 (N.Y. Sup. Ct. 1989).

[5] *Id.* at 987.

[6] *Id.*

[7] The *Castro* court identified some of the potential difficulties with forensic DNA testing. For example, the biological evidentiary sample is often of limited supply, meaning that there is no way to redo an experiment. Also, forensic samples may contain contaminants, such as bacterial DNA, that can interfere with testing procedures. Third, it may not be possible to run as many controls for forensic tests as for other tests. *Id.* at 993.

been answered affirmatively by almost every jurisdiction that has examined the problem.

Prong three asks, "Did the testing laboratory perform the accepted scientific techniques in analyzing the forensic samples in this particular case?"[8] The *Castro* court saw this as the most important inquiry, noting that a test which includes only the first two prongs may obscure critical problems in the *use* of particular forensic techniques. Because the state's testing laboratory failed in several ways to perform the accepted scientific techniques, the *Castro* court excluded the state's matching and probability evidence. The *Castro* court also suggested detailed procedures for future courts engaging in a similar prong-three inquiry.[9]

The first Alabama Supreme Court opinion to examine DNA evidence is noteworthy for its interpretation of *Castro* and for extensively differentiating between DNA matching evidence and probability statistics. In *Ex Parte Perry*,[10] an appeal from a capital murder conviction, the Court adopted a three-prong test "substantially similar to that announced in *Castro*."[11] It interpreted prong three to require two inquiries. As in *Castro*, the testing laboratory must perform generally accepted scientific techniques without error; also, the results must be *interpreted* without error. In *Perry*, the Alabama Supreme Court did not find the state's DNA evidence inadmissible. It simply held that because of the novelty and potential impact of the evidence, the trial court should conduct an admissibility hearing to address all the issues before allowing the evidence to go to the jury.

The *Perry* case distinguished DNA matching evidence and population frequency statistics in connection with each branch of the admissibility standard. It found that prongs one and two of the *Frye* test were satisfied with respect to DNA matching evidence, but that neither prong was satisfied with respect to DNA population frequency statistics. The court also held that insufficient evidence was presented to determine whether prong three was satisfied with regard to either type of evidence.[12] Moreover, even if the evidence were found on remand to be otherwise

---

[8]*Id.* at 987. As *Castro* and a number of other cases have noted, the traditional *Frye* test is satisfied by the first two inquiries. The important issue is not that a "three-prong" jurisdiction considers prong three to be part of its *Frye* test, but that the jurisdiction requires an affirmative answer to prong three before admitting the DNA evidence at all.

[9]*Id.* at 998–999.

[10]586 So.2d 242 (Ala. 1991). (The defendant Perry's lawyers never requested a hearing outside the presence of the jury to challenge the admissibility of the DNA evidence, and the matching evidence was admitted without objection. Over the objection of the defense, the state's expert testimony was also admitted. The expert testified that the DNA tests constituted a match and that the probability of finding similar DNA was 209.1 million to one.)

[11]*Id.* at 250.

[12]The court was concerned partly because the only testimony on the issue was by experts from the laboratory that performed the tests, who had obvious reasons to be biased.

admissible, the court was concerned that juries faced with enormous probability statistics, such as 209.1 million to one, would focus solely on the statistics and disregard the value of other evidence. To offset this potential for prejudice, the court required that the probative value of statistical evidence be balanced against its potential for prejudice, even if the evidence were to pass the three-prong *Frye* test.

The balancing of the probative and prejudicial value of DNA statistical evidence can be tricky. It may not actually pose a significant additional burden to the admissibility of DNA evidence. The astronomical numbers that may prejudice a jury are the same numbers that express the enormous probative value of DNA evidence. It is because DNA can exclude the vast majority of the human population from matching that a DNA match with a defendant is meaningful. Probative value clearly prevailed in one of the first Alabama appellate courts to evaluate a trial court's interpretation of *Perry*. In *Yelder v. State*,[13] Yelder was appealing his convictions for three rapes, sodomy, and burglary. Identity of the perpetrator of the crimes was a key issue in the trial. "None of the three victims could positively identify her attacker. Thus, the DNA evidence became the most compelling evidence in the present case."[14] Whether the probative value (or prejudicial value) of DNA evidence actually increases when no eyewitness testimony is available to help establish identity is an interesting question. In *Yelder*, the court held that DNA statistical evidence passed the *Frye* test and that its probative value outweighed its prejudicial value.

The Massachusetts Supreme Court, however, struck a different balance, offering what is probably the harshest criticism of DNA population frequency statistics to date in *Commonwealth v. Curnin*.[15] The court was reviewing the rape conviction of defendant Curnin, who challenged the admission of inculpatory DNA matching evidence and testimony that only 1 in 59 million Caucasians would have similarly matching DNA. Curnin had challenged the admissibility of the DNA evidence by attacking the reliability of Cellmark's forensic techniques, but the trial judge ruled that such questions went not to admissibility but only to the weight to be accorded to such evidence by a jury. The supreme court disagreed. Because of the potential prejudicial effect of DNA evidence, which has "an aura of infallibility," it mandated an inquiry analogous to the *Castro* three-prong *Frye* test.

The Massachusetts Supreme Court reversed Curnin's conviction without ruling on the propriety of the forensic DNA testing procedures because it concluded, "[T]here is no demonstrated general acceptance or inherent rationality of the process by which Cellmark arrived at its conclusion that one Caucasian in 59,000,000 would have the DNA components disclosed by the test that showed an identity between the

[13] 1991 Ala. Crim. App. LEXIS 2536.

[14] 1991 Ala. Crim. App. LEXIS 2536 at *39.

[15] 565 N.E.2d 440 (Mass. 1991).

defendant's DNA and that found [in a semen stain on the victim's nightgown]."[16] The criticism of Cellmark's statistical evidence was directed primarily to unproven assumptions regarding random distribution of genetic alleles and to the use of too small a database to measure properly genetic frequencies in a population.

*Curnin* brings into view within the legal literature concerns about population frequency statistics that may be important for future cases. A public defender doing legal research with limited resources is far more likely to discover the *Curnin* opinion than to discover scientific articles that raise the same criticisms. Still, the value of *Curnin* as precedent may be limited because the prosecution did not present an expert at trial to substantiate its probability claims.[17] The supreme court acknowledged that by the time Curnin was retried, the prosecution might be able to support the admissibility of its probability evidence.

*Jurisdictions Employing the Traditional Two-prong* Frye *Test*

A number of jurisdictions confronted with DNA evidence have employed the traditional two-prong *Frye* analysis for DNA admissibility.[18] In these jurisdictions, expert testimony opining that generally acceptable DNA testing procedures were not properly performed in the specific case at hand is treated as bearing on the weight that the jury should accord to the DNA evidence. Consequently, such testimony ordinarily will not be grounds for excluding the DNA evidence from trial. Jurisdictions employing the two-prong test tend to be more lenient toward the admissibility of DNA.[19] This leniency should not be surprising, since in three-prong jurisdictions, the majority of the attention has been focused solely on the third prong.

---

[16]It would not admit evidence of a DNA match without testimony of the likelihood of such a match occurring.

[17]This may be one of the rare instances where a defendant was better prepared to rebut inculpatory DNA evidence than the proponent was to present it.

[18]*See, e.g.,* State v. Lipscomb, 574 N.E.2d 1345 (Ill. App. Ct. 1991). (Appellate court affirmed sexual assault conviction and approved of admission of DNA matching evidence and 6.8 billion to 1 probability evidence.) The court concluded that DNA theory, RFLP laboratory procedures, and the various procedures used to generate statistical evidence were generally accepted in the scientific community. Any questions concerning the specific procedures used by the company or expert go to the reliability of the evidence and are properly considered by the jury in determining what weight to give to this evidence. The court did leave open the possibility that DNA evidence might be excluded if it is shown that the procedures used give an unreliable result.

[19]*See, e.g.,* State v. Davis, 814 S.W.2d 593 (Mo. 1991).

## Jurisdictions That Do Not Require a Frye Test

When courts in Iowa,[20] Ohio,[21] and Texas were confronted with DNA evidence, they rejected the *Frye* test and instead employed "standard" evidentiary rules regarding relevancy and expert testimony to determine admissibility. A Texas appellate court summarized its test by stating, "In general, expert testimony is admissible under the relevancy standard if the witness is qualified as an expert, the testimony will assist the jury, and the probative value of the testimony is not substantially outweighed by its prejudicial effect."[22] Interestingly, the court went on to explain that the reliability of scientific evidence depends on the three *Frye* factors: the validity of underlying scientific principle, the validity of the technique applying that principle, and the proper application of the technique on a particular occasion. The court also listed eleven nonexclusive factors to be considered in determining the admissibility of novel scientific evidence. Although this court's extended discussion demonstrates that the relevancy standard could theoretically be more stringent than a *Frye* inquiry, jurisdictions using the generic relevancy approach have typically been lenient in admitting DNA evidence.

Displaying even greater receptivity, courts in the state of Indiana will now admit DNA evidence in a criminal trial without a pretrial admissibility hearing for the defendant. In one of the first criminal trials where an Indiana prosecutor proffered DNA evidence, the trial court held a six week *Frye* hearing that generated 2200 pages of transcript and several dozen videocassettes.[23] The court found the DNA evidence admissible, and the defendant was convicted of murder. On appeal, the Indiana Supreme Court held "that the theory and techniques of DNA identification currently available are generally accepted in the scientific community as capable of producing reliable results."[24] The court declined to require proof that DNA tests were properly performed (i.e., prong three of *Frye*) because of "the abstruse, intensely technical nature of the procedural standards involved."[25] It held that once the trial court has ruled that a witness is qualified to give expert testimony regarding DNA analysis, subsequent evaluation of DNA evidence goes only to its weight and is for the jury to decide.

Eight days later, the Indiana Supreme Court ruled on a murder case in which the defendant was convicted in part because of DNA evidence that had not been

---

[20]State v. Brown, 470 N.W.2d 30 (Iowa, 1991).

[21]*See, e.g.*, State v. Blair, 1990 Ohio App. LEXIS 5812 (Ohio Ct. App. 1990).

[22]Trimboli v. State, 817 S.W.2d 785, 791 (Tex. Crim. App. 1991).

[23]Hopkins v. State, 579 N.E.2d 1297, 1301 (Ind. 1991).

[24]*Id.* at 1302.

[25]*Id.* at 1303.

subjected to a pretrial admissibility hearing. It explained that the *Hopkins* opinion constituted a conclusion that the theory and techniques of DNA identification currently available are capable of producing reliable results and thus are admissible as evidence. The court reiterated that once the proponent's DNA expert has been ruled qualified as a matter of law to testify as an expert regarding DNA analysis, questions as to whether the testing laboratory performed the accepted techniques in the particular case go only to the weight of the DNA evidence. Therefore, failure to grant the defendant a pretrial hearing did not mean that the evidence was erroneously admitted. These two Indiana Supreme Court cases eliminate the requirement of a pretrial hearing on the admissibility of DNA evidence in that state. The Indiana legislature has also passed a statute to that effect.[26]

## Approaches in the Federal Courts

The federal appellate courts that first analyzed DNA evidence reached differing conclusions as to the proper standard for admissibility. The Eighth Circuit first examined the issue on an appeal from a rape conviction.[27] DNA evidence obtained from a semen stain on the victim's underwear was admitted by the trial court upon the testimony of one expert that "DNA evidence" was generally accepted by the scientific community. Defendant Two Bulls objected, claiming that a more rigorous standard, such as a *Frye* test, should be applied. The circuit court agreed with Two Bulls, citing the novelty of DNA evidence and its potential for prejudice. Although it did not explicitly adopt a *Frye* test, the court said that whether a Federal Rules approach or a *Frye* test is used, the trial court must listen to experts from both sides and be satisfied that the evidence meets all three tests laid out in *Castro*. The circuit court set aside the conviction and remanded the case for an expanded pretrial hearing on the admissibility of the DNA evidence. The trial court was instructed to rule on five issues: the three issues subsumed by the *Frye* test, whether the evidence is more prejudicial than probative in this case, and whether the statistics used to determine the probability of someone else's having the same genetic characteristics are more probative than prejudicial.

In contrast to the Eighth Circuit, the Second Circuit rejected the *Frye* test when it first examined the admissibility of DNA evidence in *United States v. Jakobetz*.[28] Jakobetz's conviction for kidnapping was supported at trial by significant evidence in addition to the DNA matching evidence offered. Still, the appellate court

---

[26]IND. CODE §35-37-4-13 (1991).

[27]United States v. Two Bulls, 918 F.2d 56 (8th Cir. 1990). This decision was vacated pending the grant of a rehearing en banc, 925F.2d 1127 (1991). The rehearing was dismissed, and the case remanded for dismissal of the indictment after the death of the defendant, 1991 U.S. App. LEXIS 6840.

[28]955 F.2d 786.

recognized "how devastating" DNA probability statistics can be to a defendant's case. It stated, "Realistically, the results of such testing can be so dramatic as to become virtually dispositive on the question of identity, which often determines a defendant's guilt or innocence."[29] The DNA evidence being challenged on appeal[30] was admitted only after an eight day pretrial hearing at which five government and four defense experts testified.

Although the Second Circuit recognized that the majority of jurisdictions have adopted the *Frye* test to judge the admissibility of DNA evidence,[31] it applied its own "*Williams*" test,[32] which balances the evidence's probative value, materiality, and reliability against its tendency to mislead, prejudice, and confuse the jury.[33] Jakobetz argued that because DNA evidence carries an "aura of mystic infallibility," the *Williams* reliability standard should be the high standard of "reliable beyond a reasonable doubt." The Second Circuit disagreed. "[T]he court need not make the initial determination that the expert testimony or the evidence proffered is true before submitting the information to the jury."[34] Rather, the inquiry should be whether the testimony will assist in understanding the evidence or determining a fact in issue. The Second Circuit did not think that a jury would be so dazzled or swayed as to ignore evidence suggesting that an experiment was improperly conducted or that testing procedures have not been adequately established.

---

[29] *Id.* at 789.

[30] "The FBI concluded that the DNA profiles from the two samples constituted a 'match' and calculated that there was one chance in 300 million that DNA from the semen sample could have come from someone in the Caucasian population other than Jakobetz." *Id.* at 789. The FBI's Caucasian data base, used to calculate the probability statistics, was derived from blood samples of approximately 225 FBI agents throughout the United States. *Id.* at 793.

[31] The Second Circuit views the "two-prong" *Frye* test as the true *Frye* test and notes that the "third prong" applied by a number of jurisdictions normally applies only to the weight that the jury should accord to evidence and goes beyond the requirements traditionally demanded by *Frye*. Jackobetz, 955 F.2d at 794–795.

[32] The Second Circuit rejected the *Frye* test in 1978, holding that the Federal Rules of Evidence superseded it. United States v. Williams, 583 F.2d 1194 (2d Cir. 1978).

[33] The court stated, "Because the probativeness and materiality of most scientific evidence proffered to the jury are not usually in dispute, the *Williams* test boils down to a balancing of the reliability of the evidence against its potential negative impact on the jury." Jakobetz, 955 F.2d at 794. "Specific factors that could affect a court's determination of reliability: (1) the potential rate of error; (2) the existence and maintenance of standards; (3) the care and concern with which a scientific technique has been employed, and whether it appears to lend itself to abuse; (4) the existence of an analogous relationship with other types of scientific techniques and results that are routinely admitted into evidence; and (5) the presence of "fail-safe" characteristics or the likelihood that potential inaccuracies will redound to the defendant's benefit rather than his detriment." *Id.*

[34] *Id.* at 797.

After commending the lower court judge for his careful, exhaustive consideration of the issue,[35] the Second Circuit gave clear indications that, in future cases, such elaborate hearings on DNA evidence would not be necessary. It stated that in future cases a court could properly take judicial notice of the general acceptance of the underlying theory and the use of the FBI's specific techniques. It added the following:

> Beyond such judicial notice, the threshold for admissibility should require only a preliminary showing of reliability of the particular data to be offered, i.e., some indication of how the laboratory work was done and what analysis and assumptions underlie the probability calculations... . Affidavits should normally suffice to provide a sufficient basis for admissibility... .[36]

The court deemed the issue of whether accepted protocol was adequately followed in a specific case to go more to the weight than to the admissibility of the evidence:

> Rarely should such a factual determination be excluded from jury consideration. With adequate cautionary instructions from the trial judge, vigorous cross-examination of the government's experts, and challenging testimony from defense experts, the jury should be allowed to make its own factual determination as to whether the evidence is reliable.[37]

Whether other jurisdictions become increasingly willing to take judicial notice of the reliability of DNA theory and procedures remains to be seen.

## DNA Evidence and the Rights of the Accused in the Adversary Process

If truth-finding were the only important facet of the judicial process, challenges (appeals) of jury verdicts might only be permitted when the evidence presented at trial did not appear to support the trial verdict. However, the law places great importance on individual rights within the adversarial process.[38] The Constitution guarantees persons accused of violent crimes numerous procedural rights, including: protection from unreasonable search and seizure; the right to confront accusers; the right to a speedy trial by a jury of peers; and the right to assistance of counsel. In

---

[35]The appellate court believed that the district court's findings would satisfy the scrutiny of the *Two Bulls* or *Castro* test, in addition to the *Williams* test. *Id.* at 799.

[36]*Id.* at 799–800.

[37]*Id.* at 800.

[38]Similarly, the scientific community criticizes or rejects scientific results when the experimental procedures used to derive them are faulty.

numerous situations where DNA evidence was used to convict a defendant, the defendant challenged that conviction by alleging violations of these rights—violations related to the use of DNA evidence at trial. The following summary of case law describes some of these claims of error and their resolution.

Defendants' assertions that the taking of biological samples from their bodies for DNA analysis infringed upon a privacy right, or transgressed the right against unreasonable searches have been unsuccessful. This issue has arisen in paternity cases in civil court, where the right asserted was a right to privacy.[39] In criminal trials, defendants contended that requiring them to submit to withdrawal of blood prior to obtaining the order violated their constitutional rights because the state made no showing of "probable cause" that the blood would lead to inculpatory evidence.[40]

Criminal defendants have also alleged that DNA evidence admitted at trial violated their constitutional right to confront their accusers. For example, the defendant in an Ohio case unsuccessfully argued that the technician who personally conducted the DNA profiling tests should have been required to testify at trial about the actual testing procedures used.[41] In this case, testimony of another witness concerning the DNA evidence was admitted under the "business record" exception to the hearsay rule.

Some criminal defendants have alleged that delays related to obtaining DNA evidence resulted in a violation of their right to a speedy trial under the Sixth and Fourteenth Amendments. Such post-conviction allegations have usually failed, provided that the state was diligent in its pursuit of forensic DNA evidence and received permission from the court for delays.[42] However, a South Dakota court did

---

[39]*See* Sudwischer v. Estate of Hoffpauir, 589 So.2d 474 (La. 1991). (Sudwischer wanted to prove that she was the daughter of a deceased man by DNA analysis of herself, her mother's other biological children, and the deceased man's biological daughter. The man's daughter, who stood to inherit a greater share of the man's estate if Sudwischer failed, opposed the blood test necessary for this DNA testing. The court balanced the plaintiff's "constitutional" right to prove filiation to the deceased man with the potential violation of the defendant's privacy "expectation" from a compelled blood test. It held that the plaintiff's overriding financial and emotional interest in knowing her father's identity outweighed the defendant's financial interest in avoiding a blood test. The defendant asserted no physical or religious objections to this "minimal" invasion of privacy, and she had the ability to avoid the blood test by conceding the asserted biological relationship between the plaintiff and her deceased father, so the case was not a difficult one for the court. *Also see* Ohio v. Peyton, 1990 Ohio App. LEXIS 5864 (Ohio Ct. App. 1990). (The extraction of blood from defendant in a paternity suit is a minimal intrusion, and does not subject defendant to an unreasonable search and seizure in violation of his Fourth Amendment rights.)

[40]State v. Blair, *supra* note 21.

[41]State v. Fontenette, 1991 Ohio App. LEXIS 4366.

[42]*See, e.g.*, State v. Wimberly, 467 N.W.2d 499 (S.D. 1991); Lewis v. Henry, 400 S.E.2d 567 (W.Va. 1990).

dismiss the charges against a defendant when the state failed to file a motion for good cause delay and where much of the delay was due to the state's waiting to obtain DNA testing results.[43] In Arkansas, a court excluded the state's DNA evidence because the state had not made the evidence available to the defense in time for the defendant's expert witness to evaluate it adequately before the scheduled trial.[44]

The Fifth and Fourteenth Amendments to the Constitution guarantee that no person shall be deprived of liberty without "due process of law." Due process encompasses a number of protections for defendants, including protection against the government's destroying evidence that might exculpate the defendant. Criminal defendants have alleged violations of this right when the government destroyed biological samples that might be used to generate exculpatory DNA evidence. Frequently, the "destruction" of the evidence occurred when the state itself depleted a miniscule biological sample in performing its own forensic tests. Thus far, however, courts have only been willing to find that the state's destruction of evidence violated due process if a defendant has made a two-part showing that (1) the destroyed evidence had apparent exculpatory value prior to being destroyed, and (2) the state acted in bad faith in destroying the evidence.[45]

A related issue regarding the defense's right to present its case involves access to expert witnesses. Because of the power of DNA evidence and its novelty, defense experts are vital to stage an adequate challenge to prosecutorial use of such evidence. Indigent defendants are therefore especially in need of costly expert witness services. A Minnesota appellate court recently grappled with issues surrounding state payment for such experts. After timely request by defense

---

[43]State v. Head, 469 N.W.2d 585 (S.D. 1991).

[44]State v. Stuart, 810 S.W.2d 939 (Ark. 1991). (A continuance of the trial would normally have been a solution to this dilemma, but in this case would have precluded defendant's attorney from being able to represent defendant.)

[45]*See* United States v. Stevens, 935 F.2d 1380 (3rd Cir. 1991). (Sexual assault suspect's motion to dismiss on the ground that the government had destroyed a semen sample taken from the assault victim's mouth failed due to absence of showing of bad faith. He complained that, had he been able to perform DNA testing on the sample, he could have conclusively proven his innocence. The government used up the samples attempting to perform its own forensic tests, and claimed that not enough semen was recovered to perform a DNA test. Nothing in the record suggested that the FBI suspected the sample might contain exculpatory as opposed to incriminating evidence.)

*See also* Wenzel v. State, 815 S.W.2d 938 (Ark. 1991). (No due process violation when semen sample taken from rape victim was depleted during government's testing. Defendant demonstrated neither the requisite bad faith nor likely exculpatory value). State v. Williams, 1991 Ohio App. LEXIS 3883 (1991). (No error in admitting state's DNA evidence, even where insufficient blood remained for defense to conduct own test, since state did not lose or destroy any portion of sample and since defense was given access to test results and all recorded scientific data that resulted from state's DNA analysis.)

counsel, and an order permitting the retention of experts as necessary, the trial court had eventually ordered that only slightly more than half of the expert witness fees claimed by the defense should be reimbursed. The appellate court remanded, instructing the trial court to show why the fees claimed were unreasonable.[46]

Defendants have also challenged their convictions by alleging ineffectiveness of counsel, arguing that their trial lawyers improperly used or rebutted DNA evidence. Given the importance of resource balancing to the adversary process, this issue is particularly critical. In *Whitner v. State*,[47] defendant Whitner on appeal alleged ineffective assistance of counsel because his lawyer failed to object to the admission of DNA matching and probability evidence at trial. Whitner had been convicted of rape, aggravated sodomy, armed robbery, aggravated battery, burglary, and possession of a knife during the commission of various felonies. The evidence presented at trial showed that the defendant's DNA matched DNA from semen found on the rape victim and that the frequency of this DNA pattern in the general population was 1 in 94 million. Fingerprints, pubic hairs, and head hairs taken from the victim and the scene of the crimes also inculpated the defendant. Despite these other forms of evidence, an appellate court held that Whitner was entitled to a hearing at the trial level concerning his claim of ineffective assistance of counsel because of his trial counsel's failure to object to the introduction of the DNA evidence. This failure may have been particularly troubling to the appellate court because at the time of the initial trial, DNA evidence had not yet been ruled admissible in Georgia.

In contrast, a Missouri defendant unsuccessfully challenged the sufficiency of the evidence to support a conviction of first degree murder of his wife and also unsuccessfully alleged that his trial counsel was ineffective for failing to locate and present expert witnesses to refute the state's DNA evidence.[48] The DNA evidence was important for the murder conviction because, although evidence of a violent, bloody crime was found in the wife's car, the wife's body was never found to show that she was the victim of the crime. The hidden car was found in a storage garage that the defendant had rented. DNA from bloodstains in the car were compared to DNA from the husband and from two children who were biological progeny of the husband and missing woman. The "DNA fingerprint" of the blood from the two children had bands that matched bands from the accused husband and bands from the blood found in the car. At the same time, the DNA from the blood from the car did not match that of the accused husband. The Cellmark DNA expert for the prosecution testified that the DNA typing evidence indicated that, "Expressed in

---

[46] In re Application of Larry Lee Jobe, 477 N.W.2d 23 (Minn. App. 1991).

[47] Whitner v. State, 401 S.E.2d 318 (Ga. Ct. App. 1991).

[48] State v. Davis, 814 S.W.2d 593 (Mo., 1991).

percentages it was a 99.99+ chance that the blood samples came from the mother of the [son of the victim] and a 99.9999+ chance that the blood samples came from the mother of the [daughter]."[49] The defense attorney called no expert witnesses, either at the *Frye* hearing or at trial, to challenge the DNA evidence or its admissibility. The defendant argued that the prosecution's DNA expert did not perform (or oversee) the tests himself and was thus incompetent to testify to the nature and efficacy of those tests. He also complained that the expert did not sufficiently explain his basis for the statistical evidence. The Missouri Supreme Court stated that arguments "concerning the manner in which the tests were conducted [go] more to the credibility of the witness and the weight of the evidence, which is in the first instance a discretionary call for the trial court and ultimately for the jury."[50] The court held that admitting the expert testimony was not an abuse of discretion.

With respect to the allegation of ineffective assistance of counsel due to failure to locate and present DNA experts, the Missouri Supreme Court said that the defendant had to show that such experts existed at the time of trial, that they could have been located through reasonable investigation, and that the testimony of these witnesses would have benefited his defense. The defendant did not support his allegation by suggesting an expert who could have testified on his behalf, and his motion for relief failed. Despite the seeming harshness of this ruling, the court's pivotal reason for denying the motion was that the defendant's trial counsel had arranged separate DNA testing through Lifecodes, a different laboratory, and the results of that test, which were not offered at trial, were consistent with those of Cellmark.

Although the independent testing by the defense counsel would probably satisfy most people that the prosecution's DNA matching evidence was accurate,[51] it should be remembered that the role of the defense counsel is not satisfied by a personal belief that the prosecution's evidence is reliable. The defense counsel has a duty to present a zealous defense. One could argue that this duty requires use of experts to attack the state's DNA testing procedures even if counsel is independently convinced that the results of the state's tests are accurate. This defense attorney's failure to call any experts is especially troubling, since the Missouri court appears to place few barriers to the admissibility of DNA evidence, leaving the jury to assess the weight of the evidence and the credibility of the witnesses. The mere absence of contradictory testimony might enhance the credibility of the prosecution's witness—a witness who in this case did not even oversee the actual testing of the evidentiary samples presented.[52]

---

[49]*Id.* at 599.

[50]*Id.* at 603.

[51]The Supreme Court's opinion is not clear as to whether the Lifecodes test results support the probability statements of the prosecution's expert.

[52]The failure to procure a DNA expert to testify for the defendant was raised in at least one other

In Alabama, a man convicted of rape, Lightfoot, sued his attorney for failing to order a DNA test in an effort to exculpate Lightfoot.[53] The attorney won on summary judgment because at the time of Lightfoot's arrest and conviction in 1983–1984, DNA fingerprinting evidence had not been used in any trials. The negligence standard to which a defense attorney would be held today would certainly be higher, now that DNA evidence is becoming more common.

### Could Compelling DNA Evidence Alone Control the Outcome of a Case?

Because DNA probability statistics can be so enormous, the question arises whether DNA evidence alone should compel a certain result in a trial. The case law addressing this question is quite limited and comes largely from civil rather than criminal cases. Although the stakes in criminal trials are greater, arguably the most likely situation where DNA testing *should* be outcome-determinative is where such evidence exculpates a criminal defendant, because a criminal defendant need only raise "reasonable doubt" about guilt to escape conviction.[54] In other situations, the answer appears to depend on the jurisdiction and the type of lawsuit.

A number of paternity lawsuits suggest that a jury would be entitled to ignore compelling DNA evidence indicating that a man is the biological father of a child. In these cases, the plaintiff attempted to prove that the defendant was the biological father of a child by using human leukocyte antigen (HLA) blood testing procedures.

---

jurisdiction. In Frasier v. State, 410 S.E.2d 572 (S.C. 1991), the South Carolina Supreme Court agreed with the lower court that failure to call a DNA expert to testify for the defense was not ineffective assistance of counsel when the defense counsel vigorously cross-examined the state's DNA experts and attacked the accuracy of the DNA evidence. This holding was not dispositive, however, since defense counsel's performance was deficient in other respects.

[53]Lightfoot v. McDonald, 587 So.2d 936 (Ala. 1991). (Lightfoot was not challenging his rape conviction but was suing his defense attorney, McDonald, for malpractice.)

[54]*See, e.g.,* State v. Sylvester, 581 So.2d 361 (La. Ct. App. 1991). (Criminal charge of failure to pay child support dismissed after DNA fingerprinting report stated that it was impossible for Sylvester to be the biological father of the juvenile involved.) *But see* Yorke v. State, 556 A.2d 230 (Md. 1989). Yorke was convicted of rape but filed a motion for a new trial four years after the crimes were committed, claiming that he had discovered new evidence that showed that he was not the rapist. The new evidence was DNA fingerprinting data that indicated that Yorke's DNA did not match DNA contained in vaginal washings of the victim. Although the trial judge admitted this new evidence and found it material, the judge reasoned that the DNA from the washings could have been the victim's or her boyfriend's, with whom she had intercourse several hours before the rape. The victim had also testified that she was unsure whether the attacker had ejaculated. The judge did not believe that the DNA evidence would affect the jury verdict so he denied Yorke's motion for a new trial. By the time the appellate court was asked to review this ruling, further DNA tests had shown that the victim was the sole source of the DNA recovered from her body after the rape. It held that such results did not in any way eliminate Yorke as the rapist and would not affect the denial of his motion for new trial.)

Although HLA matching evidence is not as definitive as DNA matching evidence, the plaintiffs often had experts testify that HLA evidence indicated a more than 99.5% chance that the defendant in the case was the father of the child involved. Thus, HLA evidence, like DNA evidence, is a form of biological matching evidence that may be presented as generating compelling probability statistics. Juries have found for the defendant in paternity cases despite such HLA evidence, and the jury verdicts have been sustained numerous times on appeal.[55]

The Mississippi Supreme Court explained its affirmance of such a jury verdict by noting that HLA test results are not necessarily conclusive.

> The jury may consider the expert testimony for what they feel that it is worth, and may discard it entirely... . There is an unfortunate trend in our society to rely on science and technology to provide an easy and absolute answer to every question. For this reason, we would caution that the results of blood tests once received into evidence should not be viewed as absolutely determinative of the issue of paternity. Instead, they should be carefully weighed by the trier of fact in every case along with the other evidence before the court.
> 
> * * *
> 
> [J]ury findings will not be disturbed by the reviewing court merely because the findings are against the preponderance of the evidence, if they are supported by some evidence of probative force.
> 
> * * *
> 
> As long as a defendant in a paternity action has a right to a jury trial, and absent some statutory pronouncement, paternity test results, even those showing a high probability of paternity, cannot be conclusive as a matter of law.[56]

Other reviewing courts have noted that the probability statistics offered by plaintiffs are merely the probability that a man selected at random from the relevant population would be excluded by the test, and not the probability that the defendant is the father.[57] Because HLA testing generates probability statistics similar to those generated by DNA testing, one might predict from these HLA cases that jury acquittals made contrary to DNA evidence might also be upheld.[58]

---

[55]*See* D.B.J. v. B.G.B., 576 So.2d 537 (La. Ct. App. 1990), *U.S. Cert. denied*, 112 s.ct. 248. (Plaintiff's expert "calculated that there was a 99.60% probability that defendant was the father of R.J."); Chisolm v. Eakes, 573 So.2d 764 (Miss. 1990). ("The HLA test showed that there was a probability of 99.59649% that Dale Eakes was the father of Shauna Chisolm." "[O]ne out of every 20,000 Caucasian males would have the same test results as Eakes.") *Also see* State v. Hagen, 382 N.W.2d 556 (Minn. Ct. App. 1986); Smith v. Shaffer, 515 A.2d 527 (Pa. 1986). (Supreme Court reinstated jury verdict for defendant despite 99.99% probability evidence.)

[56]Chisolm, 573 So.2d at 767, 768.

[57]*D.B.J.*, 576 So.2d at 539.

[58]The deference accorded to jury verdicts also demonstrates the adversarial system's preference that

At least one New York trial court was extremely deferential to what it considered powerful DNA evidence.[59] The issue in that case also related to a paternity dispute, but it arose in the context of a man suing a mother for slander for having accused him of fathering her child. The court found that a DNA test indicating "a probability of paternity of 99.993%" was sufficient as a matter of law to warrant summary judgment[60] for the defendant mother. The *King* court explained that DNA probabilities in the 99+% range conclusively satisfied New York's "clear and convincing" standard for paternity actions.

The *King* opinion is remarkable not only because it is willing to dispose of a civil case upon DNA evidence alone, but also because it explicitly suggests that DNA and other biological tests should be the sole focus of paternity testing in the future:

> Having reached the level of scientific advance now available through the utilization of DNA probe testing in paternity testing, neither the courts nor the parties need continue to operate on nineteenth century standards... . In this court's opinion, the scientific advances of today have made the following general areas essentially irrelevant for determining paternity: sexual access by a married or unmarried male; credibility of the parties; the mother's marriage status and her sexual promiscuity or lack thereof; credibility of nonexpert witnesses presented by the parties; admission or statements made by either the male or female as to *who* is the father of the child.
>
> Once the DNA probe results reach into the 99+% range, it is difficult to believe a court would for *any* reason disregard the testing, if same has been determined to have been reliable and base its decision on, e.g., "credibility of the witnesses."[61]

Surely, the *King* opinion advocates the most extreme position in its embracing of DNA testing in paternity actions. Such an opinion is likely to please some and frighten others. One important aspect of "conventional" paternity actions is that mother, child, and defendant are all available for repeated testing should one side challenge the results of a test. Thus, many of the problems that arise with forensic criminal evidence are eliminated. Other critical differences, such as what lay jurors make factual determinations.

[59]King v. Tanner, 545 N.Y.S.2d 649 (N.Y. Sup Ct. 1989). (Denial of motion to renew slander suit dismissed on summary judgment; King v. Tanner, 539 N.Y.S.2d 617 (N.Y. Sup. Ct. 1989).

[60]The court will grant summary judgment to a moving party if no genuine issue of material fact remains to be determined at trial and it can rule in the case as a matter of law. In this case, the DNA evidence was deemed to be dispositive on the only "disputed" factual issue: the truth of the mother's statement that the defendant was her child's father.

[61]King, 144 545 N.Y.S.2d at 659 (emphasis in original).

consequences ensue and what standard of proof applies, also distinguish the civil and criminal contexts. Whether the *King* approach—allowing DNA evidence alone to determine the outcome—represents an aberration or the future commingling of science and law remains to be seen. What is already clear is that the stakes are high indeed.

## CONCLUSION

This review of recent case law shows that in considering challenges to DNA evidence, courts have thus far focused primarily on setting standards for admissibility. Their responses have varied. The trend has been to admit the evidence, but for several courts admissibility inquiries have become substantially more sophisticated and stringent than was typical in the early cases. Judges have been sensitized to important controversies regarding the forensic use of DNA evidence, including problems both of execution of the tests and of interpretation of the resulting evidence. Although admissibility has received the lion's share of critical attention thus far, other issues affecting individual rights and adversary resource-balancing are also importantly involved in the use of DNA evidence. These issues have thus far received less penetrating analysis. As noted above, both the balancing of resources and the allocation of expert as compared to lay decisionmaking roles are central to the fairness and adequacy of the adversary process. In particular, development of the defense's ability to use and challenge DNA evidence is of vital importance. Hopefully, as with admissibility, courts' increasing experience will deepen their awareness of the problems as well as improving the adequacy of their approaches to resolving these disputes.

# Forensic DNA in the Trial Court 1990–1992: A Brief History

## JEFFREY BAIRD

Senior Deputy Prosecuting Attorney
W554 King County Courthouse
Seattle, Washington 98104-2312

In the late 1980s and early 1990s, criminal trial lawyers in the United States found themselves talking, in the courtroom, to judges, juries, scientists, and each other, about DNA evidence of identification.[1] The context of the courtroom is the context of controversy; the language used there is the language of argument. In arguments that cited the testimony of scientists recruited by each side, DNA was extolled as the greatest advance in forensic evidence in this century on the one hand, and condemned as irresponsible hyperbole, as a concerted attack on civil liberties and sound scientific procedure on the other hand. Seeking either the admissibility or the exclusion of this evidence, advocates for either result employed the language of persuasion with skill and ardor. Like two fighters in a ring, trial lawyers on either side of the DNA question were preoccupied with attack and defense; judges, like anxious referees, were concerned with a fair fight and an outcome acceptable to the public. Neither judges nor attorneys expressed much interest in the historical forces that had conspired to bring science into such intense focus in the courtroom. They concerned themselves with winning their cases and with making case law. However, during that brief period of time, they may have been making history in forensic science.

It is difficult for a trial lawyer to express himself descriptively rather than persuasively (as an advocate). This difficulty may not be peculiar to attorneys; it seems to afflict every expert witness who steps into a courtroom. Much has been said and written about the admissibility of DNA evidence of identification in criminal trials; most of this has been written by attorneys and by scientists who have been opponents or proponents of the evidence. What follows is an effort to describe the forces that converged, during a brief period of time, in courtrooms where suspects were confronted with genetic evidence of their guilt or innocence and where judges and juries decided what this evidence meant and what it was worth.

---

[1] Unless otherwise designated, "DNA evidence of identification" and "DNA fingerprinting" are used interchangeably in this chapter to denote a procedure that employs restriction fragment length polymorphism (RFLP) technology (including electrophoresis, probe hybridization, and autoradiography) to assay variable numbers of tandem repeats (VNTRs) at loci used by the FBI (and most state laboratories) and in which the product rule is employed to determine an overall profile frequency. Although these are not the only polymorphisms assayed or techniques employed for forensic use, they are, at present, the most common and, because they afford the most discrimination, the most controversial.

*DNA on Trial: Genetic Identification and Criminal Justice*
© 1992 Cold Spring Harbor Laboratory Press 0-87969-379-7/92 $3 + 00

## DNA IN THE TRIAL COURTS 1990-1992

### Trials as Oral History

Case law — the law created by the courts of appeal in published opinions — is both an expression of current jurisprudence and its history. Precedents — prior published opinions — are acknowledged and either affirmed or overruled. Each articulation in the progress of case law turns upon a unique set of facts, which is likened to, or distinguished from, those heard in the courtroom before.

There is another kind of case law, which is unpublished and lacks the hierarchical power of precedent. These are the decisions of the trial courts, which arise from a particular case and apply only to the parties who have come together for trial. Of course, trial courts' decisions are reviewed by the courts of appeal, where they are either reversed or upheld, and where written case law is made. Although the decisions of the judge in a criminal trial have no legal force outside the courtroom where they are announced, these decisions precede all others chronologically: They are the stuff of which case law is made. The law the trial courts strive to follow and apply to the facts before them is old law; the decisions they make are brand new and will determine the course of law in the future. In an important sense, it is in the trial courts that legal history is made.

During the late 1980s and the early 1990s, trial courts in the United States struggled to accommodate a new development in forensic science to a body of case law ill-suited to the task. In the hiatus between the advent of DNA fingerprinting and appellate rulings on its admissibility, trial courts decided whether or not the jury would hear such evidence in criminal trials. They made findings about the facts, and conclusions about the law. These decisions ultimately determined the context in which DNA fingerprinting reached the courts of appeal and hence the way the law was written. When this new technology met criminal jurisprudence in courtrooms during these years, much was revealed about the underlying principles, strengths, and weaknesses of something called forensic science. The initial forum for this encounter in virtually every courtroom was the pretrial hearing, usually referred to as the "Frye hearing," in which opponents and proponents of the technology presented the testimony of expert witnesses in an effort to persuade the judge that their position was supported by good science and sound jurisprudence.

### Frye

In most states, the admissibility of "novel" scientific evidence (forensic evidence which has not been the subject of appellate court rulings) in criminal cases is governed by the Frye standard. *Frye* was a 1923 Washington D.C. case which disallowed the admissibility of polygraph (lie detector) evidence. No words concerning the admissibility of purportedly scientific evidence have been quoted as widely as these from *Frye*:

> Just when a scientific principle or discovery crosses the line between the experimental and the demonstrable stages is difficult to define. Somewhere in this twilight zone the evidential forces of the principle must be recognized and while courts will go a long way in admitting expert testimony deduced from a well-recognized scientific principle or discovery, the thing from which the deduction is made must be sufficiently established to have gained general acceptance in the particular field in which it belongs. (*Frye v. United States*, 293 F.2d at 1014.)

Ubiquitously recited by jurists around the country for the last 70 years, this standard has survived because it is undeniably sensible and inherently almost meaningless. As a result, it has served the courts quite well; generally, a satisfyingly conservative duration has intervened between the discovery of a scientific technique and its forensic application (usually, against a defendant).

Although the *Frye* standard remains the law in most states, its meaning remains elusive. This became embarrassingly obvious during the recent controversy surrounding the forensic use of DNA. What is the relevant scientific community, and how is general acceptance determined? Is the "particular field" made up of those scientists who, whether or not they have studied or employed a questioned technique, have elected to express their opinions on its admissibility? What if the scientists who speak the loudest for or against the use of the technique are those who are handsomely reimbursed for doing so? What if those who speak the loudest have gained all of their experience with the technique, not in the course of their own professional careers, but as "expert witnesses" — that is, not in the laboratory, but in the courtroom?

These questions were not resolved by the DNA controversy. Indeed, it is fair to say that they were ignored. The *Frye* standard was cited hundreds of times, and its meaning remained as elusive as ever. In the meantime, decisions were made. The question of the admissibility of DNA evidence was answered by the trial courts. While purporting to determine the opinion of the scientific community (an entity which proved increasingly elusive), trial courts throughout the country reached a consensus of their own — they admitted the evidence.

Unlike the classic paradigm of scientific consensus, reached by the demonstrable truth of a repeatable experiment, this consensus of opinion was nothing more than dozens of independent decisions pertaining to unique conditions. Each decision was reached independently of the others, as trial court decisions always are, according to the facts of the particular case and the witnesses who present them. The judicial consensus that developed regarding the admissibility of DNA evidence in criminal trials around this time might have appeared to the scientific community as nothing more than the circumstantial congruence of independent and singular events. Scientific principles can be demonstrated by experiments; legal principles (like the *Frye* standard) are inchoate truths, evolving over a period of time through numbers of unrepeatable, individual experiments: trials.

A similar consensus was demonstrated by those appellate courts who inherited the issue: This evidence will be admissible in the future. However they described the *Frye* standard (and most invoked it), an overwhelming majority of state and federal trial and appellate courts admitted DNA evidence because they decided, in the context of other testimony by experts and non-experts in criminal trials, that it was reliable.

## The Scientific Community

While judges were deciding that an extremely powerful new laboratory method was reliable enough for the courtroom, that is, resolving questions of science, scientists became engaged in a debate of their own, about the law. Scores of scientists were recruited by prosecutors and defense attorneys to testify in *Frye* hearings around the country. Eventually, the scientific community became frustrated with the adversarial system that had so easily conscripted so many scientists and, perhaps, disgusted by the bitter but lucrative process in which its members had become courtroom experts. They determined to resolve the question posed by *Frye* themselves. In an inquiry unprecedented in forensic science, scientists considered the ultimate legal question about the forensic use of DNA—its admissibility. Although it was not evident in the (inevitably) contentious atmosphere of the courtroom, scientists did reach a consensus. Most agreed that a five-locus "match" between even distant relatives at the VNTR loci used by the FBI was a rare event. Given the inherent limitations of gel electrophoresis, the "matching window" employed by the FBI and most state laboratories appeared reasonable. The process of "binning" virtually continuous data into discrete categories, and pooling data from those bins in which only a few alleles had been observed, seemed an appropriately conservative approach.

The most thorough and prestigious manifestation of the scientific community's opinion of DNA fingerprinting was a report prepared by the Committee on DNA Technology in Forensic Science, under the auspices of the National Academy of Sciences (NAS 1992). The report was released in April of 1992, after 2 years of preparation (and months of speculation in the courtroom and in the press about its conclusions). Although the defense bar would argue to the contrary (they had little choice, having proclaimed in advance that the report would vindicate their views), the report was an endorsement of the forensic use of DNA, albeit with strong recommendations concerning quality control, standardization, and oversight, and the collection of further data on which to base statistical calculations of profile frequencies. Curiously absent from the otherwise exhaustively thorough report was the subject of expert witnesses. Expert witnesses in forensic science have always displayed a wide range of accomplishments and qualifications; the *Frye* hearings on forensic DNA enlisted some of the best scientists in the country. They also attracted a rush of overpaid, underqualified, professional witnesses who found it much easier and more lucrative to represent the scientific community in the courtroom than to publish their own work (or even to receive grants). Apparently, the committee found it easier to look critically at the legal profession, the judiciary, law enforcement, and forensic laboratories than to acknowledge an embarrassing problem within their own community.

It was the power of DNA fingerprinting and its apparent claims of infallibility and absolute individuation that recruited so many scientists as courtroom experts, that galvanized the scientific community, and that ultimately yielded the NAS report. DNA fingerprinting demonstrated (and claimed) a power of identification surpassing the techniques of conventional forensic serology to such an extent that the difference became qualitative. To scientists who equated forensics with law enforcement, these claims seemed dangerous, if not presumptuous. It was clear that forensic technology could no longer be relied on to lag decades behind that employed in the best medical and research laboratories. Forensic science seemed to be coming of age, and the scientific community apparently decided to treat its little brother, if not as an adult, at least as an adolescent.

For many prosecutors, the NAS report was merely the most obvious harbinger of a welcome change in the way forensic science could be expected to operate in the future. Sample custody, or quality control, has long been the unstated weakness of forensic science. The statistical significance of its conclusions have been unchallenged. Experts commanding lucrative witness fees and little respect outside the courtroom lend new meaning to the notion that an idea may have currency, and they degrade their purported fields of expertise. Forensic science has been, for the most part, the province of law enforcement; a law enforcement mentality is ill-suited to promote unfettered scientific inquiry, debate, and continued technological improvement.

The following examples, cases that evolved from crimes committed in King County, Washington, illustrate the context in which DNA evidence of identification appeared in criminal trials in that jurisdiction.

## CASE STUDIES

*State v. Mr. X*

In a 2-month period during the summer of 1990, Mr. X killed three women in the Seattle area. All were young, attractive, and single; each was carefully arranged in an elaborate and sexually explicit pose after her death. The second and third murders were deliberately planned and carefully executed; the women were killed in their sleep, in their own homes. The first murder was more impulsive; X convinced an intoxicated woman to leave a singles bar and join him in a truck he had borrowed from a friend. He attacked the woman in the truck and raped her. He posed the body in the parking lot of a nearby restaurant and carefully cleaned the interior of the truck. Semen was recovered from the body of the victim; months later, a minute amount of blood was found in the truck, deep in the foam beneath the upholstery. X's library included biographies of a number of serial killers and a handbook on evidence published by the Department of Justice. The handbook had a chapter on conventional serology but did not mention DNA.

There were only a few sperm observed on the vaginal swab taken from the victim, and the blood found in the truck, which had been exposed to considerable heat for several summer months, was quite degraded. Neither the quantity of the sperm nor the quality of the blood was deemed sufficient for RFLP testing. At the recommendation of a local forensic scientist, the evidence was sent to a private laboratory in California for polymerase chain reaction (PCR) amplification and typing.

At that time, the only method of genetic testing using PCR that had prior forensic application was the amplification of the DQ-α region of chromosome 6, which was then assayed by a reverse dot-blot process capable of distinguishing six alleles (21 genotypes). The power of discrimination of the system was limited; it generated genotype frequencies on the order of 5%. X's DQ-α type was different from the victim's (and they had no alleles in common). Results were obtained from the sperm (it matched X's type) and from the blood (it was consistent with the victim's).

Weeks of pretrial hearings were devoted to the *Frye* question. Molecular biologists, forensic scientists, and epidemiologists, all with impressive credentials, testified. Experts for the defense described a number of potential problems with the technique including contamination, differential amplification, and sample mix-up. In addition to these substantive criticisms, the defense experts all voiced another kind of reservation: They did not believe that the judge or the jury could properly evaluate the evidence. They seemed to believe that intelligent non-scientists were incapable of distinguishing, for example, the PCR process from RFLP, or the lower power of discrimination of the former (in the manner in which it was used in this case) from that of the RFLP method. These scientists testified that the evidence should be excluded from the courtroom because it was just too difficult for non-scientists to understand. They suggested that it was too likely to be misrepresented by unscrupulous prosecutors (and apparently had no confidence in the defense attorney's ability to expose this).

DNA was extracted from only one of the two vaginal swabs obtained from the victim. The other swab was preserved for use by the defense. In what is becoming a routine practice, the defense argued at the *Frye* hearing and at trial that the DNA typing performed by the prosecution's witness was unreliable, but they took no steps to have "their" swab tested by anyone else. Public funds were available to them for this purpose, but their scientific advisor did nothing but squint uncertainly at the swab through the sealed plastic bag that contained it. X's appellate attorneys may allege that his trial lawyers committed malpractice by failing to develop exculpatory evidence, but his attorneys at trial knew full well what the results of additional testing would be. They made a tactical decision to leave the remaining swab where it remains today, in the freezer in the police department's evidence room.

The trial court ruled that the evidence would be admissible at trial. The range of issues litigated during the pretrial hearing and the court's comprehension of the scientific evidence are suggested in written findings, which include the following:

> The fact that a scientific procedure may yield a false result if not performed properly does not render it inadmissible; no scientific procedure is immune to operator error.
>
> The fact that further strides in the scientific development and use of PCR amplification and typing will be made does not render the current procedures unreliable; it is the nature of science to continually improve and refine its techniques and principles.
>
> The fact that the "power of discrimination of PCR may place a particular genotype within 5–10% of the population does not render the scientific technique unreliable: a limited power of discrimination does not impugn the accuracy or reliability of a scientific test.
>
> The fact that results of PCR may be subject to misinterpretation by unscrupulous adversaries in legal proceedings does not render the scientific principles unreliable.
>
> The various claims of contamination, allelic dropout, low power of discrimination, etc. are all arguments that go to the weight the jury should give the evidence, and not to its admissibility.

Despite its shortcomings — it fell six orders of magnitude short of the level of individuation possible with a multilocus VNTR profile — the evidence was quite useful at trial. It was not, statistically speaking, the most powerful evidence of X's guilt. On the victim's body when it was discovered were two types of fibers; some identical to the upholstery in the truck and others identical to the scraps of house carpet used as floor mats in the truck. The odds against this coincidence were never explicitly calculated, but the jury did not attribute it to chance.

X was convicted of three counts of first-degree murder. On the second two murders, which he had committed with more forethought, there was no DNA evidence (among his possessions at the time of his arrest were condoms). X was convicted on the strength of more conventional evidence of guilt: He left his hair on the bodies he posed so elaborately; he had no alibi; he passed the victims' jewelry around among his acquaintances like trophies.

The PCR evidence was powerful, not because it constituted proof beyond a reasonable doubt that X was, in the curious parlance of forensic science, the "donor" of the sperm, but because it proved that he was a liar: When the blood in the truck was discovered, X claimed that it was his own. It was the ability of the PCR evidence to convincingly exclude the defendant as the source of the blood in the truck, rather than its limited power to include him as a possible source of the semen, which was so persuasive. The power of forensic evidence to prove a defendant's lack of veracity, rather than his identity, is often overlooked.

*State v. Mr. Y*

In the summer of 1990, the body of a 53-year-old woman was discovered by her son, just inside the front door of her condominium. The victim had been punched, kicked, strangled, stabbed, and raped (although the exact sequence of these events was never conclusively determined) shortly after walking her dog early one Saturday morning. At least two dissimilar individuals had been observed outside the victim's residence around the time when she was murdered; Mr. Y resembled one of them.

After a lengthy investigation, samples of blood from five men, including Y, were sent to the FBI for RFLP typing. The police were suspicious of all five. The victim's son had lived with her for a time, and there were accounts of hostility between them. A neighbor, with a criminal record, was thought to resemble one of the men observed by witnesses near the scene. Another man, who was working on the roof of the victim's condominium complex on the morning of the murder, resembled the other man seen near the residence. When he was first contacted by the police, he denied that he had gone to work that day. An ex-boyfriend was an alcoholic and appeared emotionally unstable. None of these men had alibis for the morning of the murder; most of them had criminal records. Y was considered perhaps the most likely suspect (although he did not know the victim)[2] because he had been released from jail the night before the murder, because he lied about his whereabouts on the morning of the murder, and because he alone, of all the suspects, had refused to voluntarily provide a sample of his blood.

The FBI ran restriction fragments of DNA extracted from the male and female fractions of the vaginal swab, and from the six (including the victim's) blood controls, on a gel, and probed the membrane four times. All four autoradiographs unequivocally excluded every suspect but Y. One of these autoradiographs was declared "uninterpretable" (very poor quality). On each of the remaining three autoradiographs, the bands in the male fraction lane and the bands from Y's blood "matched" (significantly closer than the FBI's ±2.5% window). The frequency of the three-locus profile observed in both the forensic sample and in Y's blood was

---

[2]Only in retrospect, when they pondered the inevitable and unanswerable question, Why? did the prosecutors tentatively make another sort of connection, relying not on science, but on a sense of psychopathology. At the same time the victim was walking her dog early that Saturday morning, the defendant stood outside an apartment a block away, banging on the door. The woman who lived there was the defendant's ex-girlfriend. She had evicted him shortly before, because his idea of consenting intercourse did not coincide with hers. She knew he was getting out of jail the night before and providently left the state. Like the victim, she was almost old enough to be the defendant's mother. The resemblance between the two women was uncanny.

variously estimated, depending on the "race" of the population comprising the database, at between 1 in 300,000 and 1 in 3,000,000.

The *Frye* hearing lasted several weeks. The prosecution called a number of molecular biologists and epidemiologists familiar with the RFLP method, and population geneticists who had concerned themselves for years with the statistical analysis of substructure in human populations and human diversity. The defense called three witnesses, none of whom had ever produced a VNTR autoradiograph or published an article in a peer-reviewed journal on any topic remotely analogous to their courtroom testimony. Nevertheless, these witnesses had been advising judges around the country to exclude DNA fingerprinting from criminal trials as long as this evidence had been available. For the most part, they repeated criticisms of the technology raised elsewhere by their more renowned colleagues; their own thoughts on the subject were often presented in the embittered tones of unacknowledged prophets. They were well compensated for their courtroom expertise, too. All of them made considerably more money per year in expert witness fees than from their academic salaries or from any other source of income.

Their objections to the evidence included many criticisms addressed shortly thereafter in the report of the National Research Council. They attacked the FBI's laboratory procedures (ethidium bromide in the analytical gel, lack of monomorphic probes) and the validity of the statistical methods used to calculate genotype frequencies (database of inadequate size, insufficient proof of independence of alleles within and between loci to justify use of product rule). With equal fervor, these witnesses described themselves as victims of a grand scheme to suppress their opinions and named an impressive list of co-conspirators in this venture: the FBI, the Department of Justice, *Science* and a number of other prestigious journals which had apparently rejected articles submitted to them by the witnesses, a number of well-known academic institutions, any prominent scientist holding an opposite view, and, of course, the prosecution. (Note: In a *Frye* hearing conducted in this jurisdiction since the release of the NAS report, one of these same witnesses expanded this list to include the National Research Council itself.)

The trial court ruled that the DNA evidence of identification would be admissible in Y's trial. In the oral remarks that accompanied the judge's rulings, he considered the "general acceptance" test first announced in *Frye*. The court refused to defer the question of admissibility to the scientific community:

> The court doesn't think that the issue of general acceptance can be resolved by simply counting noses or conducting a survey of experts. Instead I conclude that the court must focus on whether a particular principle is used generally or regularly within the scientific community and whether there are any sound reasons for not permitting the same principle to be used in a court of law.
>
> Simply because certain scientists do not believe that a principle is reliable enough for forensic use is not relevant, in my judgement, to the issue. The

question is whether in the relevant scientific community the principle, technique, or methodology is considered reliable enough for its general application. If so, the court may, absent no other reasons for exclusion, permit the experts in particular case to base his or her opinion on the same generally accepted principles.

Other, corroborating evidence of Y's presence at the scene of the crime was presented to the jury. Hairs left on and around the body were similar to Y's (and to none of the other suspects'). One of the victim's neighbors drove past the victim's residence twice on the morning of the murder. On each occasion she observed a man standing just outside the residence. The clothes she described seeing on that man were identical to those worn by Y that morning. (Y claimed that he had been wearing something else, but a booking photograph proved unequivocally that he had been released from jail hours before wearing the clothes the witness described.) Of more interest to a jury, who had recently been schooled by both the prosecution and the defense in the application of the "product rule,"[3] was the creation by this witness of an Identi-Kit composite of the man she had seen that morning.

An Identi-Kit composite is a collection of transparent templates, each of which depicts a discrete area of an individual's face in frontal view. When the templates are superimposed, they combine to illustrate an entire face. The Identi-Kit consists of dozens of templates in each category; for example, any one of several dozen different noses can be chosen for a particular composite. A great variety of necks, chins, mouths, facial hair, eyebrows, foreheads, and hairlines are similarly available. Any given feature can be selected, and this selection is (physically) independent of any feature in another category. The numerical relationship between any particular composite and the total number of composites possible with the kit is on the same order of magnitude as that claimed to represent the chance that a person drawn at random from the population would have a VNTR profile identical to Y's.

The Identi-Kit evidence corroborated the DNA evidence, but it accomplished more than that. It also served as an example, closely aligned with common experience and accessible to common sense, of the enormous significance (statistical power) of relatively common events when they occur simultaneously. The Identi-Kit composite produced by the victim's neighbor, who had never seen the defendant before the morning of the murder, looked exactly like him.

Despite this evidence of his presence at the scene, the most telling evidence that Y had engaged in intercourse with the victim was testimony that he shared a three-locus VNTR profile with the sperm found in her body, and that this coincidence was not likely to have occurred by chance. The state was fortunate that Y chose to maintain a defense of identity (that was not my face, my hair, my clothing, my sperm) despite the best forensic evidence of identity available. By doing so, he, in effect, acknowledged that the man who had intercourse with the victim killed her;

---

[3]Statisticians told the jury that the chances of two independent events occurring simultaneously can be calculated by multiplying together their individual frequencies of occurrence.

his attorney described the killing as a "rape-murder." Hence, in this case, evidence that Y was the source of the sperm constituted proof that he committed murder.

Y could have changed his defense and testified that he had, indeed, engaged in intercourse with the victim, with her consent, and that he did not kill her (and he could have offered a number of plausible explanations for his earlier reluctance to admit this). Consent is not a defense to the crime of murder, but it is a defense to the crime of rape. Had he changed his defense, the most powerful evidence against Y would have been rendered completely irrelevant. Evidence that he shared a DNA profile with the sperm would have served only to prove that the defendant had sex with the victim — something his attorney would already have acknowledged to the jury. The state would have been forced to use other evidence to prove that he was a murderer.

Perhaps the uncertain legal climate of the time influenced Y's defense. The trial court had no appellate law to answer the question of *Frye*; a hearing had to be conducted to determine whether or not the principles and techniques of DNA fingerprinting were generally accepted in the scientific community. The defense attorney was advised by three "experts" who assured him that the DNA evidence was suspect and subject to easy challenge. The defense had a copy of what was purportedly a draft of the upcoming NAS report on DNA fingerprinting. The report was extremely hostile to the methods employed at the time by the FBI — particularly the methods used to estimate the statistical significance of a match.[4] *Science* magazine had recently published two opposing articles on the subject by four highly respected scientists (Chakraborty and Kidd 1991; Lewontin and Hartl 1991); was this not, itself, evidence of a lack of consensus in the scientific community? There were the autoradiographs themselves; one of them had been declared "uninterpretable" by the FBI, and on another, a band in the suspect's lane was slightly askew.

In light of this evidence, Y's attorney would have been understandably confident of his chances of convincing either the trial court (at the *Frye* hearing) or the jury, or the courts of appeal, that the DNA evidence in this case could not meet the general acceptance test of Frye or that it was simply unreliable in this particular case. If he succeeded in any one of these three efforts, his client would go free. His choice was clear: to contest the DNA evidence to the best of his considerable ability. Even if he failed to suppress the evidence in the trial, he could challenge it before a jury, which could acquit his client, and before the appellate court, which could overrule the trial court's ruling admitting the evidence and reverse the defendant's conviction.

The defendant became visibly involved in contesting the evidence. His attorney's dedication and resolve seemed to strengthen his confidence that the most

---

[4]Although this "draft" was admitted as evidence in the *Frye* hearing and in the trial, it bore little resemblance to the eventual report of the committee. Echoes of the "draft," written in the same breathless style employed by the defense experts in the *Frye* hearing, surfaced in a *New York Times* article (Kolata 1992) published just before the committee released its report and retracted the following day.

telling evidence of his guilt would never reach the jury (just as evidence of his violent criminal history was kept from them). It probably never occurred to him to consider admitting the truth of the evidence and admitting that he had engaged in sexual intercourse with the victim.

A parallel myopia had afflicted the police. The victim had lived in a small municipality outside Seattle. The police department there investigated an average of one murder a year. The detectives had conducted an exhaustive investigation, which ended when they sent blood from five suspects to the FBI. When they learned that Y's blood matched the forensic sample, they were elated. They were supremely confident that the evidence would be admitted by the judge and that it would be accepted without reservation by the jury, and they turned their attention to the myriad of other crimes being committed in their jurisdiction every week.

The prosecutors were not nearly so confident. They were impressed by the credentials, if not by the number (or, in some cases, the logic or the motives), of experts who criticized the FBI's methods. There were problems with the autoradiographs, and the profile was the product of only three loci. Most disturbing of all, the FBI had simply thrown away enough remaining DNA from the forensic sample to repeat the entire assay or to probe additional loci. To the prosecution, and certainly to the defense, this act appeared inexcusably arrogant and decidedly unscientific. The only rationale proffered by the Bureau was bureaucratic: "We've always done it this way, usually there isn't enough sample left to run a second test."

The prosecutors were required to defend the FBI's action in a pretrial hearing concerning the destruction of evidence. It was not the sort of courtroom proceeding that prosecutors enjoy; they prevailed, but only because the defense could not demonstrate even the possibility that the remaining sample might have led to exculpatory evidence. The defense attorney would (and did) argue to the jury that the FBI's behavior was further proof that the defense experts were right. There was a conspiracy to suppress evidence that DNA fingerprinting was unreliable. When the NAS report subsequently recommended an active professional-organization committee that would enforce laboratory standards and protocols, the prosecutors were not displeased.

The prosecutors knew how these various weaknesses might combine to influence a judge or a jury. However, one possibility concerned them more than any other. What if Y simply changed his story and admitted that he had sex with the victim? They repeatedly warned the police about this possibility and urged the detectives to continue their investigation into alternative sources of evidence implicating the defendant in the murder, not just in the intercourse. Just as Y seemed to harbor an irrational belief in his attorney's ability to keep the evidence from the jury, the police were mired in their own complacency. They were supremely confident that the state would prevail at the *Frye* hearing and certain that the jury would accept the DNA evidence without reservation. They considered requests from the prosecutors for additional investigation with bemusement; they were unshakable in their faith, transfixed by technology and by the power of unimaginably large numbers.

Y maintained to the jury that he never had intercourse with the victim, and the police were right. He was convicted of murder in the first degree.

One wonders if the result in the Y case would have been different had it come to trial a few years later. The defense attorney would perhaps not have been so confident about his ability to keep the DNA evidence from the jury; he might have wondered whether another defense — the defense of consent (consenting intercourse, no killing) — would serve his client better. One can expect that as DNA fingerprinting becomes admissible as a matter of course in criminal cases, defense attorneys and their clients will take steps to diminish its power to identify individuals who leave traces of their genomes where they commit the worst crimes of all. Some of these precautions will be taken after the crime, in the courtroom, by a plea to a lesser charge, or by creative perjury that admits the truth of the evidence but denies the crime. Some of the precautions will be taken before the crimes are ever committed. No doubt when the individuating power of fingerprints was discovered, this advance in forensic science portended an end to a burglar's anonymity. Yet shortly after fingerprint evidence was first admitted in criminal cases, intelligent and resourceful individuals discovered that the great power of the technique could be thwarted by a simple precaution: wearing gloves. Although crimes of violence are more impulsive than property offenses, those contemplating the former will know, as the public will learn through the news media and the entertainment industry, to wear condoms, to burn bloody clothing, to bury their victims where they will never be found.

DNA evidence of identification is often touted by its proponents (including prosecutors) as a great tool for the exclusion of innocent suspects. This argument is usually dismissed by opponents (including defense attorneys) of the technology, who apparently consider this observation disingenuous, as if protecting the civil liberties of individuals is not properly the province, or the concern, of prosecutors. Criminal defense attorneys (and hence those who align themselves with the defense in criminal cases) seem to discount the power of DNA to absolve the innocent suspect from suspicion because they have never had a client who was exonerated by this evidence. But the constitution protects the rights of all citizens, whether or not they have retained a lawyer. The reason so few defense attorneys are grateful to DNA technology for excluding their clients as suspects in rapes and in homicides is that the individuals exonerated by the technology usually never have to hire attorneys at all. They are never charged with a crime.

The Y case illustrates this phenomenon well. Blood from five suspects (and they were all "suspicious" individuals) was sent to the FBI. Only one of these men, Y, had to retain a defense attorney; the others were excluded by the autoradiographs. In the era before DNA fingerprinting, Y might or might not have been charged with, and convicted of, the murder of his victim. The other four men, however, whether any of them was charged with murder or not, would never have completely escaped suspicion by the community. The DNA evidence contributed to Y's conviction, and because it was offered against him, it was opposed by a defense attorney. The same evidence completely exonerated four other men. No attorney represents the interests of these men in the debate over DNA evidence and civil liberties, so perhaps it is appropriate that other public lawyers–prosecutors, do so.

## Unsolved Case; Uncharged Crime

A high school cheerleader shows up an hour early for practice one Saturday morning. She is abducted, dragged into the tall grass beside the football field, raped, and strangled. Two boys passing by the scene observe a man crouched over the girl, but assume they see a couple "making out." Later, one of the boys glances back to see a man running from the body. He sees the man's face in profile; later the boy tells the police that he will be able to identify the man if he sees him again.

Assume that, in the interval between the crime and a defendant's trial for that murder, DNA has been extracted from a vaginal swab taken from the body of the victim and from the defendant's blood. Assume that a state crime laboratory, using the laboratory and statistical methods employed by the FBI in the early 1990s, finds a five-locus match between the DNA from these two sources and calculates that the odds of finding another match in an individual selected at random from the general population exceeds one in a million.

Regardless of the prescribed standard that governs the admissibility of this evidence, a trial judge will inevitably compare it to other testimony that may be proffered in the courtroom by either party. Some of this will be from "lay" witnesses—witnesses who will describe "facts" within their personal knowledge. Other testimony will come from "experts"—witnesses who, by virtue of training, education, or experience in a particular field, are allowed to express "opinions." The diverse array of this evidence constitutes the context in which the admissibility of forensic DNA appears in the trial court.

The boy who saw the killer fleeing from the scene of the crime will be allowed to testify that the defendant was or was not that man. If the boy employs the language of confidence, this evidence will be presented as a statement of absolute certainty: That is (or is not) the man I saw. No calculations will be required to support the unstated equation: I am absolutely sure, 100% certain, that he, and no one else in the world, is the man I saw.

Psychologists may be allowed to tell the jury that the boy's identification is particularly reliable, or unreliable, citing "scientific" studies and anecdotal evidence in support of their contradictory positions.

If the defendant's cell-mate decides to "snitch him off," the cell-mate will be allowed to recount the defendant's alleged admission to a jury. The cell-mate's bias against the defendant may be elicited from other inmates, who may then be cross-examined about their bias against the cell-mate.

The suspect's mother will be allowed to testify that he was at home with her, as always, on that Saturday morning.

A psychiatrist called by the defendant will be permitted to venture his opinion that the suspect could not possibly distinguish right from wrong at the time of the murder. The state will be able to hire a psychiatrist who will testify that the opposite conclusion is equally evident.

Forensic odontologists—"bite-mark experts"—will be allowed to testify. One may declare that a crescent-shaped injury on the victim's chest is a bite-mark that matches the defendant's dentition. Another may testify that the defendant's teeth could not have left the injury, and a third may state unequivocally that the injury is not a bite-mark at all.

It is unlikely, but not inconceivable or unprecedented, that all of this evidence would be admitted in a single trial. Yet virtually all of it would be admissible, if a few "foundational requirements" are met. These prerequisites to admissibility would, in most instances, be trivial; a showing that the dentists and psychiatrists were licensed in the state, and, perhaps, that they could claim some modicum of forensic experience. The only consolation to the party against whom the evidence is offered is the right to cross-examine and the right to present an opposing expert. It is in this adversarial context, against this backdrop of diverse, conflicting, and contested evidence, that courts weigh arguments against the very admissibility of forensic DNA.

## CONCLUSION

During the late 1980s and early 1990s, trial courts wrestled with the question of whether DNA fingerprinting should be admissible in criminal trials. Now the question has been answered, as definitively as the question of whether it is unlikely that two unrelated people share a five-locus genotype at the loci used for DNA fingerprinting. In the future, this evidence will be admissible as a matter of course, albeit with continual improvements in accuracy and reliability. (Of course its admissibility in a particular case will have to be resolved according to those particular facts.) Each advance in technology will be litigated before it can go to the jury.

The march of forensic technology will not stop with DNA. Other means of identification, inconceivably powerful, will be developed and offered as evidence in criminal trials. Will the individuals who converge in the courtrooms with the new technology—scientists, the police, lawyers, judges—demonstrate corresponding advances?

Will the police remember that good detective work, the kind that cannot be accomplished by a telephone call to the Automated Fingerprint Identification System or the DNA databank, is still a requirement for a successful identification? Will they remember that no forensic evidence is any good at all unless it is recovered from the crime scene, sometimes by sifting for hours through apparently meaningless debris? Will they remember that many crime scenes will never be found at all unless they talk (and listen) to the human witnesses who live on the periphery of the crime?

Will the scientific community acknowledge that forensic science is an independent and legitimate field of inquiry that contributes both to public safety and to the general body of scientific knowledge? Can the scientific community insist that forensic science remain accountable to the highest principles of intellectual honesty, without forgetting that forensic science is a discrete discipline, with its own legitimate experts and expertise? Will the scientific community begin, even informally, to acknowledge that expert witness fees have created, to paraphrase one judge, a welfare system for second-rate academics?

Will forensic scientists reaffirm their first allegiance to the principles of scientific inquiry and the highest laboratory standards? Will they welcome the opinions of true experts in the disciplines where new forensic technology was originally developed? Can forensic science maintain an independent voice when its greatest source of funding is law enforcement?

Will the legislatures find ways to use new forensic technology to protect its citizens, not only from violent crime, but also from invasions of privacy?

Will lawyers find ways to express the great powers and limitations of forensic technology in everyday language? Can lawyers and jurists look ahead, from the hearings on DNA fingerprinting, to the challenges the law will face from forensic technology in the future; will they strive to imbue the principles of admissibility for such evidence with meaning, rather than honoring them as platitudes? Can lawyers look backward, from the hearings on DNA, at forensic evidence that has long been admitted without serious challenge, and begin to hold this old evidence to new standards of reliability?

The answers to these questions await the turns of history. In the meantime, a few things are certain. Even as forensic DNA technology advances, it will inevitably be overshadowed by unimaginably sensitive, discriminating, and intrusive technical means of detection and individuation. Crimes of great and inexplicable violence will go on, and the people who commit them will usually be apprehended, not by advances in science, but because they will somehow incriminate themselves, often for the same dark motives that incline them to violence against others.

## ACKNOWLEDGMENT

Special thanks are due to Robin Fox, without whom this project would not have been possible. The author is an attorney employed by the King County Prosecuting Attorney, in Seattle, Washington, where he specializes in the prosecution of cases involving homicide and sexual assault. The views expressed herein should not be attributed to his employer.

## REFERENCES

Chakraborty, R. and K.K. Kidd. 1991. The utility of DNA typing in forensic work. *Science* **254:** 1735.

Kolata, G. 1992. Chief says panel backs courts' use of a genetic test. In *The New York Times*, April 15, p. 1.
Lewontin, R.C. and D.L. Hartl. 1991. Population genetics in forensic DNA typing. *Science* **254:** 1745.
National Academy of Sciences (NAS). 1992. DNA technology in forensic science. In *Committee on DNA Technology in Forensic Science Report*. National Academy Press, Washington, D.C.

# Reliability of Statistical Estimates in Forensic DNA Typing

BRUCE BUDOWLE,[1] KEITH L. MONSON,[1] AND JAMES R. WOOLEY[2]

[1]Forensic Science Research and Training Center
Laboratory Division, FBI Academy
Quantico, Virginia 22135
[2]Organized Crime Strike Force Division
United States Attorney's Office, Northern District of Ohio
Cleveland, Ohio 44114

## INTRODUCTION

In forensic DNA identity testing, the scientist compares profiles from evidentiary samples to determine whether or not they are similar to those obtained from known samples from a suspect or victim. There are three general interpretations that can be made when performing DNA profiling: (1) inclusion, or match—the DNA profiles from the two samples are sufficiently similar and could have originated from the same source; (2) exclusion—the DNA profiles are dissimilar and could not have originated from the same source; and (3) inconclusive—there are insufficient data to render an interpretation.

If the interpretation is an exclusion or inconclusive, the process of DNA typing in the particular comparison stops; however, when a DNA profile obtained from a forensic specimen matches that of a suspect or victim, it is desirable to evaluate the likelihood of such an occurrence. Accordingly, statistics derived from population data are applied with the intent to provide the trier of fact with a guideline or estimate of how common or rare a DNA profile is in the relevant population(s) (Budowle et al. 1991a; Budowle and Stafford 1991a,b). Generally, this quantitative assessment is based on the frequency of occurrence of alleles (or more precisely, bins) (Budowle et al. 1991a) from major population categories such as Caucasians, African Americans, and Hispanics (Budowle et al. 1991b). Once the individual bin frequencies are determined, they are multiplied together applying the principle of Hardy-Weinberg (HW) expectations and gametic phase equilibrium.

Although there have been criticisms regarding the validity of population statistics applied to DNA profiling, ranging from the reliability of the technology to the validity of HW assumptions (Lander 1989, 1991; Cohen 1990; Lewontin and Hartl 1991, 1992), the scientific disagreement has distilled down to an issue of subpopulations (Chakraborty and Kidd 1991; Lewontin and Hartl 1991). The contention of some critics is that subpopulations exist in the United States, potentially resulting in large differences in allele frequencies among geographic regions. Therefore, it would be necessary to assess the frequencies of a variety of ancestral ethnic groups before estimates of the likelihood of occurrence of a DNA profile could be made with confidence. Then, based on the ethnic background of the suspect, a more precise statistic could be applied. The following discussion demonstrates that for forensic purposes, this approach is groundless and improper. The topics that are germane to the reliability and validity of population statistical approaches for DNA profiling in forensics are discussed in this paper, including the legal question to be answered, the statistical approach for assigning a likelihood of occurrence to a DNA profile, the existence and effect of subpopulations, the contributions to human variation, variations revealed by protein genetic marker data, the appropriate reference database, and forensic significance of misassignment of a reference database.

## SCIENTIFIC DISAGREEMENT

The cause of the controversy regarding the application of population statistics to DNA profiling is a difference in emphasis between two scientific disciplines. Population substructure is a phenomenon of interest to human population geneticists who search for subtle differences between populations so as to define the structure and ethnohistory of the human race. In contrast, forensic scientists deal with an unknown situation where specific ethnicity is difficult to define, and, therefore, they attempt to apply statistical inferences as a guideline or estimate on the rarity of a particular DNA profile based on readily defined groups. The FBI (Budowle et al. 1991a) and others (Evett and Gill 1991; Evett and Pinchin 1991) always have recognized that subgroups exist, but they have realized that for practical purposes the forensic significance is minimal.

Accordingly, the forensic community employs conservative statistical approaches to minimize possible differences due to subpopulations, and the differences that remain are of little forensic significance (Chakraborty and Kidd 1991; Evett and Gill 1991; Evett and Pinchin 1991; K.L. Monson and B. Budowle, in prep.). There is no evidence to support the contention that an individual will receive unfair bias, regardless of the general database used. On the contrary, it has been demonstrated that similar estimates of rarity are obtained, regardless of the database used (K.L. Monson and B. Budowle, in prep.). Basically, the human population geneticist and the forensic scientist are interested in contrasting issues relating to genetic profile data; there is little disagreement about basic assumptions of population genetics. Before the scientific disagreement is resolved, the proper defining of the question the forensic and judicial communities seek to answer should be addressed.

## LEGAL ISSUE

What is the germane legal question that is answered by DNA-typing statistics? Practitioners of forensic science must be mindful of certain aspects of the criminal justice system. These include the constitutional presumption of innocence enjoyed by all defendants, the legal definition of relevant evidence, and the nature of cases in which DNA testing is likely to be used. When these factors are taken into consideration, it becomes evident that the subpopulation concern is not particularly meaningful in the context of the use of DNA-typing statistics in criminal cases.

When a crime, such as a sexual assault or a homicide, is committed, the police typically question the victim and/or potential witnesses to attempt to ascertain a description of the perpetrator (i.e., the source of the DNA in the crime scene sample). Since ethnic origin, unlike major racial origin (Lewontin 1972), is not manifested readily by distinguishable characteristics, a witness rarely, if ever, is able to provide specific and accurate ethnic information regarding the actual perpetrator. Therefore, if the police are able to ascertain anything at all about the physical makeup of the true perpetrator, it could be his or her race. (Of course, there can be many situations where there are no eyewitness accounts or the reliability of the account is questionable.)

If a suspect, when arrested, denies guilt on grounds other than self-defense or consent, the suspect is claiming that he or she is not the source of the biological sample found at the crime scene. Therefore, someone else must be the source of the crime scene sample. Under the Fifth and Fourteenth Amendments of the Constitution, the suspect is presumed innocent, so the suspect's claim of not contributing the sample is presumptively valid. Thus, the legal question becomes, What is the likelihood that someone *other than the suspect* was the source of the evidentiary material? Bearing in mind that the ethnicity of the class of people, other than the suspect, who could have left the sample is rarely if ever known by the police or victim or witnesses, the answer to the question of appropriate database must be phrased in terms of some sort of general population group.

Under the circumstances described above, the relative rareness of a DNA pattern in the suspect's ethnic subgroup, which may be of academic interest, is not particularly relevant in the legal setting. Relevant evidence, as defined in Federal Rule of Evidence 401, is evidence having any tendency to make the existence of any fact that is of consequence to the determination of guilt or innocence more or less probable than it would be without the evidence. The relative rareness of the DNA profile in the suspect's ethnic subgroup (or in any ethnic subgroup, for that matter) is not legally relevant under this definition. It does not tell the jury anything about the likelihood that someone other than the suspect could have, in fact, left the sample at the crime scene. Instead, it only tells the jury the likelihood that someone in the suspect's ethnic subgroup could have left the crime scene sample. This has no bearing on the question of guilt or innocence in the typical criminal case. The relative rareness of the pattern in some general population of potential perpetrators, on the other hand, does help the jury assess the likelihood that someone other than the defendant could have left the crime scene sample, and this has a direct bearing on the question of guilt or innocence.

In addition to the lack of relevance of the suspect's ethnicity in the typical case, there are other legal considerations that undermine the importance of the suggestion that the rareness of the pattern in the suspect's subgroup must be considered. How do the police ever determine the suspect's true ethnic makeup? Suspects may not know their precise ethnic makeup and, even if they know it, they may rely on the Fifth Amendment right to remain silent and refuse to disclose it. Suspects may misrepresent their ethnic makeup, just as suspects in our criminal justice system frequently use aliases. On whom do we rely for this information? Should it be a lawyer, a friend, or a relative? It is obvious that determination of ethnicity of the suspect generally is not possible.

Regardless, some critics (Lewontin and Hartl 1991) have suggested that a wide variety of ancestral populations should be analyzed first because the allele frequency values in these populations have some legal bearing. The following hypothetical case, based on a realistic courtroom setting, illustrates the lack of relevance of the relative rareness of a DNA pattern in the suspect's ethnic subgroup. A woman is raped in Los Angeles. A red-headed man is seen fleeing the scene. The victim is able to describe her attacker in terms of his race, his size, and his hair color. A red-headed suspect is arrested. Through testimony, the suspect claims his ancestors are from a town in Ireland where red hair is very common. The suspect also claims he did not commit the crime and therefore another red-headed man must have committed the crime. In its effort to assess the significance of the red hair in terms of the issue of guilt versus innocence, can the jury in any way be aided by the knowledge that red hair is very common in the suspect's home town in Ireland? Obviously, this bit of ethnic information (true or not) has no bearing on the particular case in Los Angeles. Instead, the jury needs to assess the likelihood that someone else with red hair could have committed the crime in Los Angeles, and the rareness of red hair in the general Caucasian population would certainly assist them in this regard. Substitute the term "DNA profile" for the term "red hair" in this hypothetical case—the red hair described by the victim and the witness is the DNA crime scene pattern and the red hair on the suspect's head is the suspect's DNA pattern—and the lack of relevance of ancestral ethnic differences, or subpopulations, in the typical DNA case, has just been described.

Finally, in discussing the legal question that is answered by DNA-typing statistics, it also should be stressed that the same DNA evidence can be used to answer different questions that may arise in the atypical case in which some relevant subgroup is identified. On the basis of the facts of a particular case, there may be some legal relevance to the likelihood that a relative of the suspect could have contributed the crime scene sample. It is entirely appropriate in that case to tell the jury that the chance of a brother's having a four-locus profile which occurs in the general Caucasian population at 1/30,000,000 could be 1/256 and the chance of an identical twin's carrying the same type is 1/1. All are appropriate descriptions of the DNA evidence, based on the facts of a particular case, that may or may not be relevant.

The important point here is that the assessment of whether there is a relevant subgroup in a particular case is a legal decision to be made by judges, after hearing legal arguments from attorneys, based on an analysis of the evidence in that case. It is not a decision to be made by scientists. The scientist's role is to provide an answer to the general question (i.e., What is the likelihood that someone other than the defendant could have left the crime scene sample?) that arises in the general case. If there are unusual circumstances which dictate that a different legal question must be asked in a particular case (e.g., What is the likelihood that the suspect's brother could have left the sample?), it is the function of the lawyers and the legal system to frame that question based on a legal assessment of what is or is not relevant in that particular case.

In summary, the legal question is, What is the likelihood that someone from the general population other than the defendant could have contributed the crime scene sample? The question is not, What is the likelihood that someone from the suspect's ethnic subgroup (or for that matter any ethnic subgroup) could be the source of the biological material? The answer to the latter question is not legally relevant in the typical criminal case. The FBI's approach to answering the proper legal question is described below.

## FBI STATISTICAL APPROACH

The FBI advocates a fixed-bin approach for estimating DNA band frequencies. DNA fragment lengths, derived from DNA profiles of individuals of a population group sample, are measured. The measured values are categorized by class limits or "bins" defined by fixed boundaries. The boundaries must be wider than the measurement error of the analytical system. The number of DNA fragments residing in each bin divided by the total number of chromosomes in the sample population determines the frequency of a bin. Bins with less than five observations are merged with adjacent bins to establish a minimum acceptable bin frequency (Budowle et al. 1991a). During DNA profiling in a case, the frequencies of the measured bands from the DNA profile are assessed by determining in which bins they reside after adjustment for measurement error. If the measurement error range spans a bin boundary, the value assigned to a band is that of the higher frequency bin (Budowle et al. 1991a).

The fixed-bin method was employed as an operational means to compensate for the quasi-continuous nature of variable number of tandem repeat (VNTR) alleles and, by grouping different alleles together, to provide conservative estimates of the likelihood of occurrence of DNA profiles in the relevant population(s). It also provides an overestimate for previously unobserved alleles that were not detected because of practical limitations of population sample size and the highly polymorphic nature of VNTR loci. For DNA typing, the likelihood of occurrence of a multilocus DNA profile is determined by the multiplication rule. Allele frequencies within a locus are multiplied by applying the square law (Hardy 1908) to obtain single-locus genotype frequencies, and these genotype frequencies are multiplied to provide an estimate of a multilocus genotype probability.

## SUBGROUP ISSUE

The reference population for bin frequencies is categorized into, for example, Caucasian, African American, southeastern Hispanic, and southwestern Hispanic (Balazs et al. 1989; Budowle et al. 1991b; Gill et al. 1991). The major population categories, instead of further ethnic subdivisions, were chosen for databases because they are more readily defined, generally provide the largest frequency differences among groups, provide good approximations for frequency estimates for forensic purposes in the United States, and are the most appropriate in a legal and forensic context (Budowle and Stafford 1991a,b). Some critics have argued that none of these groups is genetically homogeneous, and each is composed of various ethnic subgroups; thus, if the allele frequencies of VNTR loci are noticeably different among the subgroups, the assumption of allelic independence for applying the multiplication rule to arrive at a frequency estimate would not be precisely correct (Cohen 1990; Lewontin and Hartl 1991). However, for this to become a concern, allele frequency differences must be large (Chakraborty and Kidd 1991; Risch and Devlin 1992), and the final multilocus frequency estimate from different databases must differ sufficiently to be forensically significant.

It is known that small amounts of gene flow among groups tend to minimize differences in variation among subgroups. It is intuitively obvious that gene flow among ethnic groups within a racial group (e.g., French, English, and German) is a more likely occurrence than between racial groups (e.g., Caucasians and Black Africans), such that subgroups within a race should be more similar than are the racial groups. For example, United States Caucasians derive from different European groups who, after immigration, experienced substantial gene flow. Kennedy (1944) showed that the percentage of out-marriages by national origins of Europeans during 1870, 1900, 1930, and 1940 was 9%, 25%, 35%, and 37%, respectively, and after 1900, out-marriages by religious affiliation exceeded 10% (Chakraborty and Jin 1992). All this was based on a society that was not as mobile as a post-World War II United States; it is expected that gene flow increased substantially after World War II.

Lewontin and Hartl (1991) have suggested that another element to consider for maintaining endogamy in the United States (ethnic subgroup pockets in the United States) is geographical distance. They cite a study by Spuhler and Clark (1961) that demonstrated that one third of marriages are between people born less than 10 miles apart. According to Lewontin and Hartl (1991), this mating pattern would help maintain a biologically subdivided United States population. In actuality, the data suggest the opposite. As of 1961, the majority of marriages (two thirds) were contracted by people born *more* than 10 miles apart. Thus, in a highly mobile society, marriage partners are considerably less constricted by locality. This Spuhler and Clark (1961) study is further evidence of substantial gene flow in the United States. In addition, a 10-mile distance should not have a major effect on maintaining endogamy. For example,Washington, D.C. (or any other major city) would contain many ethnic groups within 10 miles of each other. Thus, although distance could have some effect, the number of ethnic groups within the allotted distance provides for ample genetic mixing. Alternatively, Evett and Gill (1991) considered the practical implications of the effects of subgroups by simulating an extreme, unrealistic example of population stratification. Allelic data for D2S44 from Caucasians and Afro-Caribbeans were pooled, but to exaggerate the differences between the groups, 2000 base pairs were artificially added to all the allele sizes in the Caucasian samples. Thus, a database file was created exhibiting significant population substructure, greater than could ever be anticipated. The construct had little significance on evaluation of the likelihood of occurrence of a DNA profile from the pooled sample when compared with the constituent groups.

## HUMAN VARIATION

Additional support for the appropriateness of general population groups in the United States as relevant databases is found in human diversity studies. Lewontin (1972) demonstrated that the major contribution of total human variation (approximately 85%) is due to individual variation and that ethnic and racial differences are minor contributing components. Lewontin (1972) concluded that

> It is clear that our perception of relatively large differences between human races and subgroups, as compared to the variation within these groups, is indeed a biased perception and that, based on randomly chosen genetic differences, human races and populations are remarkably similar to each other, with the largest part by far of human variation being accounted for by the differences between individuals. Human racial classification is of no social value and is positively destructive of social and human relations. Since such racial classification is now seen to be of virtually no genetic or taxonomic significance either, no justification can be offered for its continuance.

The logical extension of this conclusion would be to do away with the racial group classification and even to pool all data into one group. It would appear that not much is to be gained for forensic purposes by further subdivisions of major population groups. General United States population groups as reference databases are consistent with Lewontin's recommendations.

## PROTEIN GENETIC MARKER DATA

Part of the experience of the forensic science community regarding the practical implications of subpopulations on frequency estimates for forensic purposes is based on protein marker data. The protein marker data indicate that for forensic purposes, subpopulations have little effect on estimation of likelihood of occurrence of a genetic marker profile. Surprisingly, Lewontin and Hartl (1991) suggest that differences among United States ancestral ethnic groups can have a significant impact on multi-genotype frequency estimates. They cite selected frequency data from Mourant et al. (1976) for the ABO, Rh, and Kell blood groups from Poles and Italians. Putting aside the issue of the information content and quality of protein genetic marker data derived from the 1940s and early 1950s (Budowle and Stafford 1991b; Chakraborty and Kidd 1992), the impact of the blood group marker example of Poles and Italians is better addressed in the context of all available data in Mourant et al. (1976). Both Budowle and Stafford (1991b) and Chakraborty and Kidd (1991) demonstrate that the blood group data and their differences with regard to allele frequencies do not support the contention that subpopulations will ultimately affect an estimate of the occurrence of a particular genetic profile in a forensic case. Endogamous groups (e.g., a neighborhood predominately Polish or Italian) could exist to a degree in some places in the United States. However, the Polish and Italian data cited by Mourant et al. (1976) should be evaluated in proper perspective for the forensic context. It should be expected that a predominately Polish (or Italian) neighborhood in the United States will be composed of people whose ancestors are not from one city or geographic area of the ancestral country but from various regions of the original country. Lewontin and Hartl (1991) used blood group examples from specific regions, and the regions were of different origin for each genetic marker. This mosaic construct of a population (K.L. Monson and B. Budowle, in prep.) would hardly represent any ancestral group or portray the situation of populations in the United States. An average of Kell, ABO, and Rh allele frequencies for all Polish and Italian data might logically serve as a more meaningful estimate of genetic marker frequencies for potential endogamous groups in the United States. When evaluated in this manner, the differences in frequency estimates for the blood group markers between Poles and Italians are not substantial (Budowle and Stafford 1991b; Chakraborty and Kidd 1991; Sokal et al. 1991). Furthermore, ethnic groups in the United States will intermarry and decrease any preexisting differences. These effects were observed without the benefits of placing statistical buffers on the allele frequencies of the genetic markers to compensate for potential differences between population subgroups. If conservative estimates are applied to these values, such as used with the fixed-bin approach for DNA profiling, there should be no significant differences in a forensic context for estimating genotype frequencies.

## HARDY-WEINBERG EXPECTATIONS AND CONSIDERATIONS

An HW population is a theoretical population that meets the criteria of (1) no selection, (2) no emigration, (3) no immigration, (4) no mutation, and (5) large size. Clearly, no human population satisfies the criteria of an HW population, but for the purposes of estimating the occurrence of a DNA profile in a relevant population, it matters little whether or not a particular population meets HW criteria. On the other hand, it is important that the particular genetic markers meet HW expectations, i.e., that the alleles associate independently so genotype frequencies can be estimated. The data for protein genetic markers and restriction-fragment-length polymorphism (RFLP) DNA markers indicate that for the vast majority of genetic markers in human populations, HW expectations are satisfied (Chakraborty and Kidd 1991).

Most scientists in the field of forensic science, and probably most human geneticists, accept the multiplication rule, i.e., allelic independence, as an operating assumption for forensic purposes. Regardless, some critics have suggested that there are significant deviations from HW expectations in the form of excess homozygotes in the population databases and that these deviations are due to population substructure (Lander 1989). However, the apparent excess of homozygotes has been shown to be a technical artifact of the RFLP analytical system. Single-band DNA patterns cannot unequivocally be deemed homozygotes: They could be heterozygotes where either a smaller-sized allele goes undetected (Budowle et al. 1991a; Chakraborty et al. 1992; Chakraborty and Jin 1992) or the two alleles of the heterozygote type are not completely resolved (Devlin et al. 1990). Without taking this phenomenon into consideration, excess homozygosity would be artificially inflated.

Chakraborty and Jin (1992) have shown that observed HW deviations could not be due solely to subpopulation differences. In fact, the level of excess homozygosity based on single-band patterns in United States Caucasian and African American population samples was inconsistent with the ethnohistory of these groups. Chakraborty (1991) concluded that it would require 20–30 subgroups experiencing no gene flow for 40,000 years for the observed levels of excess homozygosity to be due to population stratification. This represents an impossible scenario, since, for example, European groups could not have differentiated from each other more than 20,000–25,000 years ago (Nei and Roychoudhury 1982). Further genetic mixing of groups, particularly common in the United States, would make population substructure an unlikely cause of HW deviations. However, this observed excess homozygosity is consistent with some single-band patterns being pseudo-homozygotes resulting from a class of null alleles (Chakraborty et al. 1992; Chakraborty and Jin 1992).

Weir (1992) further elucidated the phenomenon of single-band patterns in a database. He recognized that single-band patterns can represent either homozygotes or heterozygotes. Where tests for associations based on a likelihood ratio test were confined to confirmed heterozygote patterns, no associations between alleles could be found for Caucasians or African Americans.

## REFERENCE DATABASE AND FORENSIC SIGNIFICANCE

Since it is evident that the use of ethnically subdivided databases would have little impact on the estimation of the likelihood of occurrence of a DNA profile, either scientifically or in a legal context, the only questions remaining are (1) what major population group database should be used as the relevant database in a forensic case and (2) whether there is any evidence of allele frequency differences among various databases, consisting of major population groups in the United States, that would have a forensically significant impact on the trier of fact. The answer to the first issue, based on the constitutional presumption of innocence afforded the defendant and the unknown history of the sample, is that it is not up to the DNA typer to decide which group should be used as the relevant database(s). All available general databases should be provided, e.g., Caucasians, African Americans, Hispanics, and the trier of fact can make a decision regarding the most pertinent database(s) and values based on all relevant information pertaining to the case. As to the second issue, generally it is irrelevant which database is employed. Chakraborty and Kidd (1991), K.L. Monson and B. Budowle (in prep.), Risch and Devlin (1992), and Weir (1992) have shown that multilocus DNA profiles are rare occurrences regardless of the database in the United States, whether defined by population group or by geography. Often between major population groups there can be estimates, for example, of likelihood of a multilocus DNA profile of 1/50,000,000 versus 1/5,000,000, or 1/30,000,000 versus 1/42,000,000. Examples of this magnitude would not alter the opinion of the trier of fact regarding the rarity of a DNA profile. Never has there been an example, because of the polymorphic nature of the currently employed DNA markers, where estimates of 1/50,000,000 versus 1/100 were obtained for relevant United States populations. Again, there will be no anticipated bias in a forensic context.

## CONCLUSION

Statistical methods for estimating the likelihood of occurrence of DNA profiles are robust. Reliable and valid estimates can be obtained for forensic purposes based on general population databases. Because of the rarity of DNA types in all relevant databases, there should be no bias of any forensic significance afforded an accused individual.

## ACKNOWLEDGMENT

This is publication number 92-07 of the Laboratory Division of the Federal Bureau of Investigation. Names of commercial manufacturers are provided for identification only and inclusion does not imply endorsement by the Federal Bureau of Investigation.

## REFERENCES

Balazs, I., M. Baird, M. Clyne, and E. Meade. 1989. Human population genetic studies of five hypervariable DNA loci. *Am. J. Hum. Genet.* **44:** 182.

Budowle, B. and J. Stafford. 1991a. Response to expert report by D.L. Hartl submitted in the case of United States versus Yee. *Crime Lab. Dig.* **18:** 101.

———. 1991b. Response to "population genetic problems in the forensic use of DNA profiles" by R.C. Lewontin submitted in the case of United States versus Yee. *Crime Lab. Dig.* **18:** 109.

Budowle, B., A.M. Giusti, J.S. Waye, F.S. Baechtel, R.M. Fourney, D.E. Adams, L.A. Presley, H.A. Deadman, and K.L. Monson. 1991a. Fixed-bin analysis for statistical evaluation of continuous distributions of allelic data from VNTR loci, for use in forensic comparisons. *Am. J. Hum. Genet.* **48:** 841.

Budowle, B., K.L. Monson, K. Anoe, F.S. Baechtel, D. Bergman et al. 1991b. A preliminary report on binned general population data on six VNTR loci in Caucasians, Blacks, and Hispanics from the United States. *Crime Lab. Dig.* **18:** 9.

Chakraborty, R. 1991. Statistical interpretation of DNA typing data. *Am. J. Hum. Genet.* **49:** 895.

Chakraborty, R. and L. Jin. 1992. Heterozygote deficiency, population substructure and their implications in DNA fingerprinting. *Hum. Genet.* **88:** 267.

Chakraborty, R. and K.K. Kidd. 1991. The utility of DNA typing in forensic work. *Science* **254:** 1735.

Chakraborty, R., M. de Andrade, S.P. Daiger, and B. Budowle. 1992. Apparent heterozygote deficiencies observed in DNA typing data and their implications in forensic applications. *Ann. Hum. Genet.* **56:** 45.

Cohen, J.E. 1990. DNA fingerprinting for forensic identification: Potential effects on data interpretation of subpopulation heterogeneity and band number variability. *Am. J. Hum. Genet.* **46:** 358.

Devlin, B., N. Risch, and K. Roeder. 1990. No excess of homozygosity at loci used for DNA fingerprinting. *Science* **249:** 1416.

Evett, I. and P. Gill. 1991. A discussion of the robustness of methods for assessing the evidential value of DNA single locus profiles in crime investigations. *Electrophoresis* **12:** 226.

Evett, I. and R. Pinchin. 1991. DNA single locus profiles: Tests for the robustness of statistical procedures within the context of forensic science. *Int. J. Leg. Med.* **104:** 267.

Gill, P., S. Woodroffe, J.E. Lygo, and E.S. Millican. 1991. Population genetics of four hypervariable loci. *Int. J. Leg. Med.* **104:** 221.

Hardy, G. 1908. Mendelian proportions in a mixed population. *Science* **28:** 49.

Kennedy, R.J.R. 1944. Single or triple melting pot? Intermarriage trends in New Haven. 1870–1940. *Am. J. Sociol.* **49**: 331.

Lander, E. 1989. DNA fingerprinting on trial. *Nature* **339**: 501.

———. 1991. Invited editorial: Research on DNA typing catching up with courtroom application. *Am. J. Hum. Genet.* **48**: 819.

Lewontin, R.C. 1972. The apportionment of human diversity. *Evol. Biol.* **6**: 381.

Lewontin, R.C. and D.L. Hartl. 1991. Population genetics in forensic DNA typing. *Science* **254**: 1754.

———. 1992. Response. *Science* **255**: 1054.

Mourant, A.E., A.C. Kopec, and V. Domaniewska-Sobcak. 1976. *The distribution of the human blood groups and other polymorphisms.* Oxford University Press, New York.

Nei, M. and A.K. Roychoudhury. 1982. Genetic relationship and evolution of human races. *Evol. Biol.* **14**: 1.

Risch, N.J. and B. Devlin. 1992. On the probability of matching DNA fingerprinting. *Science* **255**: 717.

Sokal, R.R., N.L. Oden, and C. Wilson. 1991. Genetic evidence for the spread of agriculture in Europe by demic diffusion. *Nature* **351**: 143.

Spuhler, J.N. and P.J. Clark. 1961. Migration into the human breeding population of Ann Harbor, Michigan, 1900–1950. *Hum. Biol.* **33**: 223.

Weir, B. 1992. Independence of VNTR alleles defined as fixed bins. *Genetics* **130**: 873.

# Statistical Issues in DNA Identification

**DONALD A. BERRY**

Institute of Statistics & Decision Sciences
and Comprehensive Cancer Center
Duke University, Durham, North Carolina 27706

## I. INTRODUCTION

A man is charged with rape and murder. A suspect is found. DNA "profiles" in a semen sample from the victim's body are compared with those in blood of the suspect. An expert says the "odds are a hundred billion to one that the two samples are from the same individual." Is the suspect guilty? On the basis of the expert's statement, most juries would say "yes," but the expert's statement is wrong! It is *impossible* to calculate *any* odds that two samples are from the same individual based only on the DNA profiles. Even if the statement is repaired by giving a correct interpretation to the hundred billion to one figure, there are uncertainties that render the magnitude of such a figure questionable.

The expert's statement is a statistical inference, since it involves probabilities. Statistical inferences using DNA profiling in cases of criminal identification are similar in most respects to those using other kinds of genetic evidence. Such inferences use characteristics of two or more samples of human body tissue to address the question of whether two individuals are identical. This question cannot be answered with certainty, at least not with existing technology. This is so even if the DNA profiling laboratory's testing is flawless. In practice, there is always the possibility of laboratory testing errors. Still another source of error is statistical variation in the laboratory's database, which is a sample from a population of interest.

In Section II, I show why the expert's statement above is wrong. Sections III, IV, and V consider questions motivated by the example of Section II. In Section VI, I outline the process by which DNA profiling using restriction-fragment-length polymorphism (RFLP) analysis is currently used in criminal cases. Sections VII and VIII describe a better method.

## II. ABO BLOOD TYPE AND BAYES' RULE

This section introduces Bayes' rule, which is the fundamental tool for learning under uncertainty. I use a specific though rather easy example.

Blood is found at the scene of a crime. Suppose temporarily and for simplicity that this is the criminal's blood, although this could never be known with certainty. A laboratory carries out a test for ABO blood grouping and finds the crime sample to be blood type AB. A suspect is identified whose blood type is also found to be AB. Only 5% of the individuals in the laboratory's database are type AB. How should this information be used in trying the suspect? In particular, what does it mean about the probability of the suspect's guilt?

Suppose for the sake of argument that the laboratory's testing is error-free. Suppose also that the database is a random sample from an appropriate population (more on these issues later). Take the criminal's blood type as background evidence: I will take it to be understood in every probability calculation. Let E stand for the evidence that the suspect is type AB and let G be the event that the suspect is guilty. The complement of G is ~G, so ~G means the suspect is innocent. We want the probability of guilt given evidence E. The standard notation for this *conditional probability* is P(G|E). We know P(E|G), the *inverse probability*: Since we are assuming an error-free laboratory process, P(E|G) = 100%.

Probabilities P(G|E) and P(E|G) are related by *Bayes' rule:*

$$P(G|E) = \frac{P(E|G) \times P(G)}{P(E|G) \times P(G) + P(E|\sim G) \times P(\sim G)}$$

In the example at hand, P(E|~G) = 5%. Therefore,

$$P(G|E) = \frac{100\% \times P(G)}{100\% \times P(G) + 5\% \times P(\sim G)}$$

Written in terms of odds rather than probabilities, Bayes' rule says that the odds in favor of guilt are

$$\frac{P(G|E)}{P(\sim G|E)} = \frac{P(E|G)}{P(\sim E|G)} \frac{P(G)}{P(\sim G)}$$

In words, the *posterior odds* in favor of guilt equal the product of the *likelihood ratio* P(E|G)/P(~E|G) and the *prior odds* in favor of guilt, P(G)/P(~G). The likelihood ratio is the probability of the evidence assuming the suspect is guilty to the probability of the evidence assuming the suspect is not guilty. In the example at hand, the likelihood ratio is 100%/5%, and so

$$\frac{P(G|E)}{P(\sim G|E)} = 20 \times \frac{P(G)}{P(\sim G)}$$

Since we know the likelihood ratio, the remaining requirement for calculating P(G|E) is P(G); P(~G) is just 1 − P(G). This is the *unconditional* or *prior* probability of guilt — "prior" in the sense that it is based on evidence other than E. To illustrate, suppose P(G) = 50%. Then the prior odds in favor of guilt are 1:1, or 50:50. The posterior probability of guilt is 20/21 = 95.2%. The posterior odds are 20:1 — the same as the likelihood ratio. The following table shows the relationship between the prior and posterior probabilities and also between the corresponding prior and posterior odds:

| P(G) | P(G\|E) | Odds (G) | Odds (G\|E) |
|---|---|---|---|
| 1% | 16.8% | 1:99 | 20:99 |
| 10% | 70.0% | 1:9 | 20:9 |
| 50% | 95.2% | 1:1 | 20:1 |
| 90% | 99.4% | 9:1 | 180:1 |
| 99% | 99.95% | 99:1 | 1980:1 |

The ABO blood type in this example provides relatively strong evidence in favor of guilt: The likelihood ratio of 20 is the largest when both crime sample and suspect are type AB. This is because AB is the rarest blood type. Suppose both crime and suspect samples had been blood type A, and suppose the population proportion of type A is 45% (typical of many populations). Then the likelihood ratio in favor of guilt would be only one ninth (= 0.05/0.45) as large as 20: namely, 100/45, or about 2.22.

Bayes' rule makes it clear that there are two quite separate components of the posterior probability of guilt; one is the likelihood ratio and the other is the prior probability. I address these in the next two sections.

## III. LIKELIHOOD RATIO

For a particular piece of evidence, the likelihood ratio of guilt to innocence is the ratio of the probability of the evidence given guilt to the probability of the evidence given innocent. In the example of the previous section, the likelihood ratio is 1/0.05 = 20. This means that evidence E is 20 times more likely under G than under ~G. It most assuredly *does not mean* that the suspect is 20 times more likely to be guilty than innocent. The latter is a statement about posterior probability, and such statements cannot be made separate from assumptions about prior probability, i.e., separate from other evidence in the case.

It is difficult to overemphasize this last point. Scientists and lay people confuse the likelihood ratio with a statement about posterior probability. My experience suggests that it is almost impossible for people *not* to be confused. An example is the following seemingly innocuous account of DNA profiling from the *Minneapolis Star Tribune* (December 20, 1991): "If the two samples give the same results when tested, advocates of the method say, it means the odds are overwhelming that one person was the source of both samples." (Compare this with the expert's statement in the introduction.) This is reversing the conditional. It may be what "advocates of the method say," but as I have indicated in the previous section, this is most assuredly not what "it means."

Here is a correct version of the quote in the previous paragraph: "If the two samples give the same results when tested, this is much more likely when the samples were from one person than when there were two different sources." This sentence is stilted, and I do not know how to make it more readable without making it wrong. (Journalists hate stilted sentences. I know a journalist — not the one quoted in the previous paragraph — who was willing to say that the likelihood ratio was the posterior odds even knowing that this was incorrect! She protested to my suggestion that she not say something she knew to be wrong: "Dr. Berry, my readers are scientists, not statisticians!")

Reversing the conditional and regarding the likelihood ratio to be a posterior odds is sometimes called the *prosecutor's fallacy*. Based on the inability of humans to avoid it, I suspect that most jurors fall prey to it, with or without help from the prosecution. The likelihood ratio is not the odds in favor of guilt, and there is no way to convert it into the odds of guilt without accounting for the prior odds of guilt. In the example, 20:1 is *not* the odds in favor of guilt. Depending on the prior probability, the odds in favor of guilt can be as small as 0 (certainly innocent) and as large as $\infty$ (certainly guilty). These two extremes occur when the evidence separate from the ABO blood typing is conclusive to the person assessing it.

For any particular set of assumptions about the reliability of the tests and about the population frequencies, the likelihood ratio is a complete representation of the genetic information. In particular, once the likelihood ratio of guilt to innocence is known, no other aspect of the evidence is relevant. (However, the *assumptions* under which the likelihood ratio is calculated can and should be questioned.)

## IV. PRIOR PROBABILITY OF GUILT

How should genetic evidence be communicated to a jury? As is evident from the previous section, there is no universal probability of a suspect's guilt. Any such probability depends on the person being assessed. The appropriate assessors in a criminal case are the jurors. Prior probability assessments can be made as described by Berry (1990, 1991); the sense of "prior" here is that it is separate from the genetic evidence. One juror might have a prior probability of guilt of 90%, whereas another assesses it to be 10%. Regardless of the genetic evidence, their posterior odds of guilt will differ by a factor of 81, which is 90/10 divided by 10/90.

Communicating quantitative evidence to jurors who have prior probabilities is easy: Apply Bayes' rule. But not all jurors will be willing to specify a prior probability of guilt. Because juries and witnesses cannot carry out a dialogue in court, it is not possible to tell whether and which jurors have assessed their prior probabilities. It may be safe to say that none have. Furthermore, those who have, may have done it badly. Thus, perhaps prior probabilities should never be addressed in a court.

Not mentioning prior probabilities, however, is much more dangerous than causing confusion by introducing them. This is so even if jurors are presented with correct statements about quantitative evidence, say, "The evidence is 20 times more likely if these two individuals are identical than if they are not." In Sections II and III, I discussed the tendency of people to reverse the conditional in this statement and think of the 20:1 as posterior odds. A juror who thinks of this as "These two samples are 20 times more likely to be from the same individual than not" is taking the prior probability of identity to be 50%. Imputing a particular and necessarily arbitrary prior probability is wrong, as is making calculations and statements that will assuredly be misunderstood. Although it is difficult to accomplish, informing jurors about the role of prior probabilities in *any* probability statement is essential.

There are many aspects of a case that can contribute to a juror's prior probability of guilt. These include all the "usual" types of evidence: eyewitness testimony, alibis, etc. An aspect that may not be regarded as important is now crucial: the manner in which the suspect came to be a suspect. Consider two cases. In both, a man is stabbed in a struggle and blood is found at the crime scene. The blood is determined not to be the victim's. In the first case, the victim's business partner has a motive and becomes a suspect. In the other, there is no suspect in the usual sense. A search carried out in a large genetic database turns up a match, someone who has no known connection with the victim and who lives in another state. The prior probability of guilt in the first case should be much larger than in the second.

The second case in the previous paragraph points up a real difficulty in probability assessment. When a suspect is identified by searching through a database, the prior probability of guilt should be very small. It may be impossible for a juror to distinguish between small probabilities such as one in a million and one in a billion. The distinction is important, since the ratio of these probabilities is 1000. Suppose the likelihood ratio is one billion — not unusually large when using DNA profiling. If the prior probability is one in a million, then the posterior probability of guilt is 99.9%. When the prior probability of guilt is one in a billion, the posterior probability is only 50%. I do not know how to deal with the problem of discriminating among prior probabilities that are very small.

A problem with updating knowledge in a legal case via Bayes' rule (or in any other way!) is that evidence may well be used twice. Suppose a suspect is identified from a database search and the case comes to trial. What should be a juror's probability of guilt before hearing any of the genetic evidence? Clearly, had the genetic evidence excluded the "suspect," he would not have been brought to trial — in fact, he would never have been identified! The juror knows this. The man *was* brought to trial. A juror might therefore assign a moderately large prior probability of guilt. However, this uses the genetic evidence implicitly. When the evidence is introduced in court, it is not appropriate to use such a prior probability in updating via Bayes' rule — this probability is not really "prior" to the genetic evidence! It is not clear how to account for such implicit duplication, but either the prior or likelihood ratio (or both) should be reduced in some way.

## V. COMBINING EVIDENCE FROM MULTIPLE GENETIC SYSTEMS

Suppose in the example of Section II that data are available on both crime and suspect samples concerning Rh factor as well as ABO blood groupings. For both genetic systems, the suspect's phenotypes agree with those in the crime sample. The population proportions of the observed (but unspecified) genetic characteristics are shown in the table below. This table also includes the likelihood ratio for each system.

| System | P(phenotype\|~G) | Likelihood ratio |
|---|---|---|
| ABO | 5% | 20 |
| Rh | 10% | 10 |
| Overall | | 200 |

For this table, I assumed the ABO phenotype is AB. The Rh system is more complicated. I have not specified the Rh phenotype, but I assume it is present in 10% of the population. The overall likelihood ratio of 200 is calculated by multiplying the individual system likelihood ratios.

The overall likelihood ratio can be used in Bayes' rule just as an individual likelihood ratio. For example, if $P(G) = 50\%$ then

$$P(G|E) = \frac{200}{201} = 99.5\%$$

Another way of viewing the above calculation of the likelihood ratio is as follows. Assuming independence, these are the proportions of the database that match on the respective systems:

| Matches crime sample? | Proportion |
|---|---|
| ABO yes, Rh yes | 0.5% |
| ABO yes, Rh no | 4.5% |
| ABO no, Rh yes | 5% |
| ABO no, Rh no | 90% |

The only relevant row of this table is the first. Since the first proportion is 0.5%, the likelihood ratio is $1/0.5\% = 200$.

Still another way to update is to incorporate the individual likelihood ratios one at a time. Starting with $P(G) = 50\%$ and using only $E_1 = $ (ABO phenotype) gives $P(G|E_1) = 20/21$. Then incorporating $E_2 = $ (Rh phenotype) using Bayes' rule gives

$$P(G|E_1 \text{ and } E_2) = \frac{P(G|E_1)P(E_2|G \text{ and } E_1)}{P(G|E_1)P(E_2|G \text{ and } E_1) + P(\sim G|E_1)P(E_2|\sim G \text{ and } E_1)}$$

In the present example,

$$P(G|E_1 \text{ and } E_2) = \frac{(20/21) \times 100\%}{(20/21) \times 100\% + (1/21) \times 10\%} = \frac{200}{201}$$

the same as in the previous paragraph in which $E_1$ and $E_2$ are incorporated together.

I evaluated the factor $P(E_2|\sim G \text{ and } E_1)$ in the above expression as 10%. This is the same as $P(E_2|\sim G)$. That these two probabilities are the same means that I am assuming that $E_1$ and $E_2$ are *independent* (given $\sim G$).

Are genetic systems independent? I know of only one study (Grunbaum et al. 1978) that addresses this question for the blood systems discussed above. This study actually addresses *pairwise* independence, which is a special case of independence. The study considered data from four ethnic groups, the largest sample (Caucasians) containing 6004 individuals. The authors found very little evidence of pairwise dependence, which suggests that multiplying any two of the individual likelihood ratios may be reasonable, but does not mean that likelihood ratios for more than two systems can be multiplied.

Suppose genetic systems are dependent, or *linked*, and a laboratory assumes they are independent. The likelihood ratio they calculate is then wrong; but is it too large or too small? For reasons I will not go into, there is a tendency for it to be too large, i.e., for it to overestimate the evidence, but it may be too small as well. Consider two artificial extremes in the ABO/Rh example given above. In one, suppose every member of the population in question who has type AB blood also has the Rh phenotype. (This means the remaining 5% of the individuals with this phenotype are distributed among the other three blood types.) These are the database proportions:

| Matches crime sample? | Proportion |
|---|---|
| ABO yes, Rh yes | 5% |
| ABO yes, Rh no | 0% |
| ABO no, Rh yes | 5% |
| ABO no, Rh no | 90% |

Given that the suspect is type AB, the fact that his Rh phenotype matches the crime sample's is irrelevant: The combined likelihood ratio should be 20 and not 200. On the other hand, suppose only a small fraction, say 0.1%, of individuals in the database with AB blood have the Rh factor in question. These are then the proportions:

| Matches crime sample? | Proportion |
|---|---|
| ABO yes, Rh yes | 0.1% |
| ABO yes, Rh no | 4.9% |
| ABO no, Rh yes | 5% |
| ABO no, Rh no | 90% |

The correct likelihood ratio is 20/0.001 = 20,000, which is up substantially from 200.

The calculations of the previous paragraph suggest a way around assuming independence. Instead of multiplying proportions for ABO and Rh phenotypes, calculate the *joint* proportions for these phenotypes. The laboratory simply finds the proportion of individuals in the database who match the crime sample at *both* systems. For example, suppose 1% of the database agrees with both the ABO and Rh phenotypes of the crime sample. This means that the database proportions are as follows:

| Matches crime sample? | Proportion |
|---|---|
| ABO yes, Rh yes | 1% |
| ABO yes, Rh no | 4% |
| ABO no, Rh yes | 9% |
| ABO no, Rh no | 86% |

The likelihood ratio is 1/0.01 = 100. Since this is smaller than 200, these data evince a form of positive dependence between the ABO and Rh phenotypes: 20% [= 1%/5%] of those who match on ABO also match on Rh as compared with only 10% overall matches on Rh.

This is a fine program since it obviates the need for assuming that genetic systems are independent, and it is easy to explain to a jury: "The suspect matches the crime sample and 1% of our database similarly matches." However, a problem develops as the number of genetic systems used increases, and therefore the genetic typing becomes more specific. Namely, fewer and fewer individuals in the database will match the crime sample on all systems. For example, suppose a database contains 2000 individuals and none of them matches. What to conclude? Obviously, a match is pretty rare, but quantifying it is not easy. Is it rarer than 1 in 2000? After all, the next two individuals added to the database may match. The answer involves statistics again! Simply reporting to a court that no one in a database of 2000 matches sidesteps the issue of independence and may be an appropriate compromise.

## VI. DNA TECHNOLOGIES

The principal DNA technology used by laboratories today is restriction-fragment-length polymorphism (RFLP) analysis, also called variable number of tandem repeats (VNTR) analysis. (For a detailed description, see Office of Technology Assessment 1990.) Technicians measure fragment lengths of DNA from both crime and suspect samples. If the individuals are the same, then the lengths of DNA fragments from both will be identical. Laboratories use regions of DNA that are "polymorphic," so the lengths will likely be different if the samples are from unrelated individuals. However, two fragments that have different but similar lengths may not be distinguished by the measuring process. From the point of view of inference, the difference between using RFLPs and ABO, for example, is that the former provides measurements on a (roughly) continuous scale and the latter contains a finite number of categories and so is discrete.

The other major DNA profiling technology is polymerase chain reaction (PCR). The genetic characteristics measured using PCR are (essentially) discrete and so are not subject to the same considerations of measurement error as are RFLPs. This means that the same type of analysis used in the ABO system (Section II) applies for PCR as well. In the remainder of this chapter, I focus on RFLPs. In this section, I describe the standard "match/binning" analysis of RFLP evidence in the context of the famous case of *New York v. Castro*. I then describe a more appropriate analysis, one keyed to the actual measurement error distributions of the fragment length measurements.

On February 5, 1987, Vilma Ponce and her 2-year-old daughter Natasha were stabbed to death in their Bronx, New York apartment. Ponce's common-law husband suggested to police that a neighbor, a 38-year-old Hispanic named José Castro, was the murderer. Investigators discovered a bloodstain on Castro's watch. They sent the watch along with blood samples from Castro and Vilma and Natasha Ponce to Lifecodes Corporation in Valhalla, New York for RFLP analysis. Lifecodes concluded that the lengths of four fragments of the blood on the watch matched those of Vilma Ponce's DNA, but that they did not match Castro's. They reported that the frequency of such patterns in the general Hispanic population was only 1:189,200,000.

In August, 1989, the judge in the Castro case ruled that Lifecodes' DNA analysis was inadmissible as evidence at Castro's trial (except for testimony that the blood on Castro's watch was not his own). The judge's ruling was confined to the procedures followed by Lifecodes in this case and did not apply to RFLP analysis generally.

Lifecodes reported fragment sizes for four RFLPs. The evidence in two of these was problematic (Lander 1989). Figure 1 shows the band weights (fragment lengths, in base pairs) for the other two RFLPs, which corresponded to loci D17S79 and D2S44.

Two bands were visible in both the watch and Ponce lanes for loci D17S79. Only one band was visible in both lanes for loci D2S44. According to Lander, given the limited amount of DNA material extracted from the stain on the watch, there is some question as to whether a larger-sized band would have shown up in its lane had one been present. This was a controversial issue at the Frye hearing in Castro. I will not address matters related to this important question, and I assume both

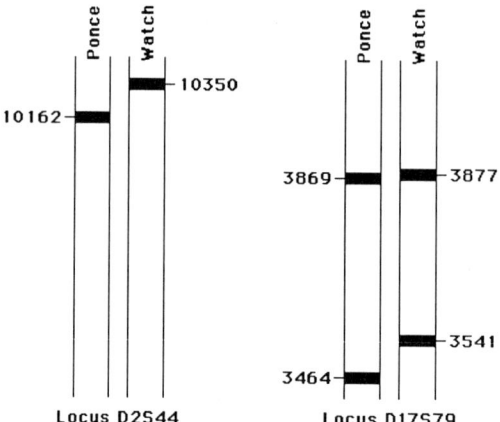

**Figure 1**
Band weights measured by Lifecodes Corporation in *New York v. Castro*; two of four loci used. Cited by Lander (1989).

samples were homozygous at this locus — or rather, measurably homozygous in the sense that the sample's two bands did not resolve separately.

Laboratories report RFLP analyses using what is called match/binning: First, they say whether there is a "match" in the two samples, and if there is, they calculate a population match frequency for the crime sample's DNA fragments. They use a formal definition in the latter and treat the first rather casually, perhaps deciding on a match visually. *They should use the same definition for both.* They may claim a visual match between the suspect's and crime sample's band weights even though the suspect's band weights would not be included in the population match frequency.

Calculating a match proportion requires an estimate of the variability in the RFLP measurement process. Using duplicate measurements of the same individuals, Lifecodes estimates the standard deviation of the difference between two measurements to be 0.6% of their average. I use this estimate, but my analysis of their data suggests that 1.0% is closer.

First consider locus D17S79. For both Ponce and the watch, there are two different bands. Consider the one for which Ponce's length is 3869. Three standard deviations is $3(0.006)(3869) = 70$ (base pairs). Lifecodes used a sample of 295 Hispanics as a reference population (see Fig. 2). (I criticize the use of this population in Section IX.) The proportion of their Hispanic sample who have fragment lengths between $3869 \pm 70$ (or 3799 to 3939) base pairs is about 0.111 (see Fig. 2). (For the Castro case, Lifecodes used an obscure method not based on a number of standard deviations and calculated a matching frequency of 0.023. Eric Lander showed that their method was defective, and my understanding is that Lifecodes later dropped it

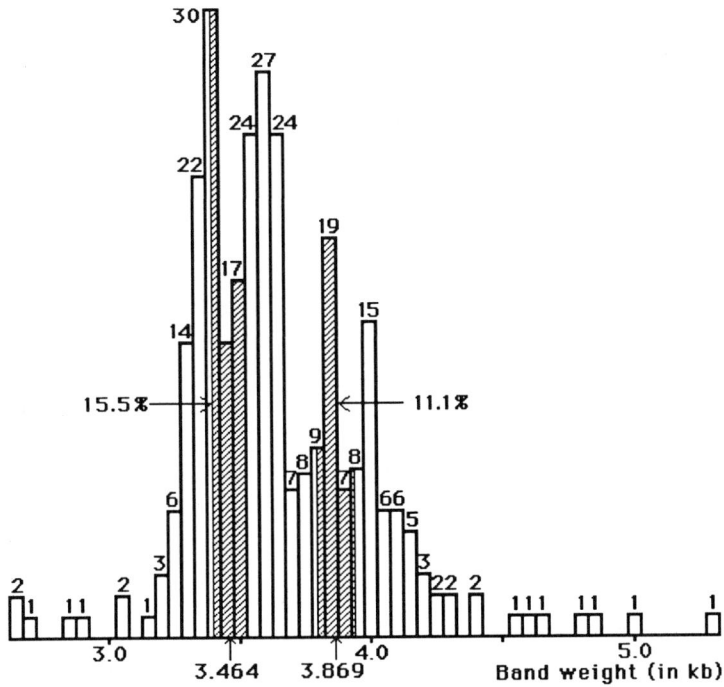

**Figure 2**
Frequency distribution of band weights for locus D17S79; 295 Hispanics. (Adapted from Fig. 1C of Balazs et al. 1989.)

in favor of ± 3 S.D.) For the 3464 band, 3 S.D. is about 62. The proportion of the population between 3402 and 3526 base pairs is about 0.155 (see Fig. 2).

The FBI uses a slight variant of this approach. For each locus, they partition the scale of possible band weights to form "fixed bins." When a band weight falls into a particular bin, that bin's relative frequency is used as the estimate of the proportion of the population that matches that band. Neither the fixed bins of the FBI nor the "floating" bins used by Lifecodes and others are clearly better. Both have advantages and disadvantages, but they are more similar than different, and both are inferior to the approach described in Sections VII and VIII.

According to Bayes' rule,

$$\frac{P(G|E)}{P(\sim G|E)} = \frac{P(E|G)}{P(\sim E|G)} \cdot \frac{P(G)}{P(\sim G)}$$

The genetic evidence affects the likelihood ratio, $P(G|E)/P(\sim E|G)$. Take Ponce's band weights to be background evidence — it is taken as given throughout. Evidence E is that the band weights on the bloodstain are within 3 S.D. of 3869 and 3462.

Castro can be guilty (event G) even if the blood on his watch is not that of Ponce (or her daughter). On the other hand, he can be innocent even if the blood stain is Ponce's. It is straightforward to use Bayes' rule to account for these two possibilities, but both numerator and denominator of the likelihood ratio become more complicated. Let event H be that the blood on Castro's watch is Ponce's. The *law of total probability* gives

$$P(E|G) = P(E|H)P(H|G) + P(E|\sim H)P(\sim H|G)$$
$$P(E|\sim G) = P(E|H)P(H|\sim G) + P(E|\sim H)P(\sim H|\sim G)$$

where I have assumed that P(E|H and G), for example, equals P(E|H). This seems reasonable, since calculating the probability of the band weights observed in the bloodstain depends only on whose blood it is.

In these expressions, P(H|G) and P(~H|G) are the probabilities that the blood is and is not Ponce's assuming that Castro is guilty, and P(H|~G) and P(~H|~G) are the probabilities that the blood is and is not Ponce's assuming that Castro is innocent. As usual, these probabilities are the purview of the jury. For convenience and so as to avoid obscuring my main points, I assume that both P(H|G) and P(~H|~G) equal 1. (This assumption is unrealistic, but it is trivial to modify the above expressions using the actual probabilities.) Then P(E|G) = P(E|H), and P(E|~G) = P(E|~H); also, the likelihood ratio equals P(E|H)/P(E|~H).

Match/binning does not give a likelihood ratio and so it is not consistent with Bayes' rule. However, the inverse of a matching proportion is a likelihood ratio, in the following sense. Take the numerator of the likelihood ratio, P(E|H), to be 1 when the laboratory decides the samples match, and 0 when there is no match. In addition, take the denominator P(E|~H) to be the match proportion. In the Castro case, for locus D17S79 the latter is 2 × 0.111 × 0.155 = 0.0344, assuming that the Hispanic population is in Hardy-Weinberg equilibrium (Office of Technology Assessment 1990, p. 67). Assuming a match on this locus, the match/binning likelihood ratio is 1/0.0344 = 29.1. Using this interpretation, the evidence from this locus is about 29 times more likely if Castro is guilty than if he is innocent.

Now consider locus D2S44. Ponce's band weight is 10,162 base pairs. The standard deviation is 0.6% of 10,162, or about 61 base pairs. The proportion of the Lifecodes' Hispanic sample that is within 3 × 61 = 183 base pairs of Ponce's (i.e., between 9,979 and 10,345) is 0.049. Assuming both Ponce's and the watch sample's are homozygous at D2S44 (as I indicated above, this was a controversial issue at the Frye hearing in Castro), the matching proportion is $(0.049)^2 = 0.0024$.

Assuming independence of the two loci (see Section V for a discussion of independence of genetic factors), the probability of the observed band weights for loci D17S79 and D2S44 is 0.0344 × 0.0024 = 0.000083. Thus, the combined match/binning likelihood ratio is 1/0.000083, or about 12,000.

The following table gives the percentage differences:

|  |  | Locus | | |
|---|---|---|---|---|
|  |  | D2S44 | D17S79 | |
| Source: | Watch | 10350 | 3877 | 3541 |
|  | Ponce | 10162 | 3869 | 3464 |
|  | Difference | 188 | 8 | 77 |
|  |  | (1.85%) | (0.21%) | (2.22%) |

Obviously, the criterion for calculating match proportions is less stringent than ±1.8%. It is logically important to use the same criteria in deciding whether a suspect sample (or, in this case, the watch sample) matches a crime sample as in deciding what proportion of the general population matches. *Probabilities for matching criteria not used for matching suspect and crime samples are meaningless.*

Consider a criterion based on numbers of standard deviations, say ±3. Suppose seven of eight suspect's bands are within 1 S.D. of the criminal's, but the eighth is 3.1 S.D. away. It is tempting to claim a match. Then we would have to adjust the match proportion according to our de facto criterion. We have already made one change; how can we be sure what our criterion really is? Can we define what we regard to be a match sufficiently well to allow for deciding which members of the population match? Looking at each member of a database to decide visually which samples match the crime sample seems fraught with danger — and also with work!

At the other extreme, suppose all eight of the eight bands of the suspect are 2.9 S.D. from the criminal's (some at −2.9 and others at +2.9). It is a match, but it is a weak match — much weaker than in the case of the previous paragraph. It should be taken as much weaker evidence against the suspect.

Yes/no matching criteria are inappropriate when using RFLP profiling. I describe a better approach in the next section. Technical details and further results are given by Berry (1991), Berry et al. (1992), and Evett et al. (1991).

## VII. BELL-SHAPED MEASUREMENT ERRORS

The distribution of measurement errors in Home Office Forensic Science Service (United Kingdom) laboratories is shown by Berry et al. (1992; Fig. 4) to have a bell-shaped or Gaussian distribution. Data from Lifecodes Corporation (M. Baird, pers. comm.) show that error distributions of repeat measurements evince the same bell shape.

I have described likelihood ratio calculations for bell-shaped curves (Berry 1991). The evidence is stronger (likelihood ratio larger) when a suspect's band is very close to that of a crime sample. The strength of the evidence decreases as the suspect's band gets further from that of the crime sample.

Consider the Castro case. These calculations are shown for the individual band weights in the following table. The table shows the results for four different measurement standard deviations; the first is the one advertised by Lifecodes and the others are given for comparison. The rightmost column shows the overall likelihood ratios making the same assumptions as in Section VI: The bands are independent and both samples are homozygous at D2S44.

|  |  | Locus |  |  | Overall |
|---|---|---|---|---|---|
|  |  | D2S44 | D17S79 |  |  |
| Source: | Watch | 10350 | 3877 | 3541 |  |
|  | Ponce | 10162 | 3869 | 3464 |  |
| S.D. |  | likelihood ratios |  |  |  |
| 0.6% |  | 0.31 | 21.7 | 0.016 | 0.017 |
| 1.0% |  | 3.05 | 13.7 | 0.772 | 49.1 |
| 1.5% |  | 4.94 | 9.2 | 2.25 | 251 |
| 5.0% |  | 4.31 | 4.3 | 3.72 | 148 |

vs. 12,000 for match/binning

For a measurement standard deviation that is 0.6% of the band weight, the evidence in these two loci is about 60 times more likely assuming the bloodstain on Castro's watch is *not* Ponce's than assuming it is. As I have indicated previously, Lifecodes' estimate of 0.6% is wrong for the duplicate data on which it is based, and 1.0% is better. Assuming the latter, the table shows that the evidence in these two loci is about 50 times more likely assuming the bloodstain is Ponce's than assuming it is not. (Although it is irrelevant to the above discussion, it is interesting that Castro eventually pleaded guilty.)

## VIII. ACCOUNTING FOR BAND SHIFTING

Standard procedure in the match/binning approach is for laboratories, including the FBI's, to multiply match proportions for different bands on the same locus and on different loci. This is not appropriate; nor is it appropriate when assuming bell-shaped measurement error distributions to multiply the individual likelihood ratios. The reason is that band measurements are dependent. This is not to say that the band weights are themselves correlated (and in fact there is substantial evidence that they are independent [Weir 1991; I.W. Evett et al., in prep.]) only that their measurements are correlated. This generally recognized phenomenon is called band shifting (Office of Technology Assessment 1990, pp. 10–11).

A laboratory can assess the effect of band shifting by taking two samples from the same individual and processing them separately. Suppose the two measurements are plotted, with one band weight in the horizontal scale and the other in the vertical scale. Berry et al. (1992) show that the resulting distribution is well approximated by a two-dimensional bell curve. Furthermore, they give an example in which the correlation between the two band measurements is 0.90.

Consider a particular crime sample with two bands. Suppose a second measurement is taken from the same individual. This is very likely to be near the crime sample measurement, but if one band in this second measurement is smaller than the corresponding band in the first, the same relationship will usually hold for the other band. Similarly, if the difference between the first and second measurement is large for the first band, the same will usually be true for the other band.

Berry et al. (1992) consider the example shown in the following table:

| Band weights | | | |
|---|---|---|---|
| crime | Suspect A | M/B | LR |
| 2790 | 2687 | no match | 1.6 |
| 3290 | 3193 | | |

| Band weights | | | |
|---|---|---|---|
| crime | Suspect B | M/B LR | LR |
| 2790 | 2840 | 105 | 1E-15 |
| 3290 | 3204 | | |

"Suspect A" actually is a duplicate measurement of the "crime" sample, whereas "Suspect B" is a different individual. Both bands of Suspect A are less than the corresponding bands of the crime sample, and by roughly the same amount. Band 1 of Suspect B is smaller than band 1 of the crime sample, but band 2 of Suspect B is larger than band 2 of the crime sample. As a result, assuming a bell-shaped error distribution, Suspect A's likelihood ratio (LR) of identity with the crime sample is much larger than is Suspect B's, even though both of B's bands are closer to the crime sample's than is A's.

The table also shows the results of match/binning, using a ±3 S.D. match criterion. Match/binning comes to the wrong conclusion in both cases. The likelihood ratio for Suspect A is only slightly larger than one, indicating that the evidence from this locus is slightly more likely assuming guilt than assuming innocence. This may not be enough to convict, but other loci will be used and they will very likely be decisive. On the other hand, the likelihood ratio for Suspect B is 0.000000000000001. Thus, even though match/binning concludes a match, Suspect B would be effectively eliminated using this better technique. (Although this is a criticism of match/binning, I do not want to overstate it. Using other independent loci, one of them will very likely [correctly] exclude Suspect B. However unlikely, it is possible that the other loci and any other evidence available will be equivocal and Suspect B will be wrongly convicted.)

An important question is how frequently the circumstance in this example occurs. To address this, Berry et al. (1992) consider 218 samples with duplicate measurements using genetic probe YNH24 at locus D17S79. Analogous to the setting in which crime and suspect samples are from the same individual, they considered all 218 comparisons within individuals. To ascertain what occurs when the crime and suspect samples are from different individuals, they considered all 23,653 between-individual comparisons.

Berry et al. (1992) consider two independent loci, accounting for correlations for bands within the same locus but assuming independence across loci. First consider the "withins." About 3.6% are excluded by match/binning, whereas only 0.006% have likelihood ratio less than one. Moreover, in the latter cases, the likelihood ratio was not much less than one; so other loci would likely demonstrate identity. In addition, whereas both the match/binning proportion and the likelihood ratio vary, the latter tends to be about 10 times as large as the inverse of the former, which, as I have indicated, can itself be viewed as a likelihood ratio.

The benefit of assuming bell-shaped error distributions is not as great for the "betweens." Both methods correctly exclude the "suspect" in more than 99.9% of these cases. However, there *is* a benefit: The bell-curve likelihood ratio now tends to be smaller than the inverse of the match/binning match proportion.

Evett et al. (1991) show how to calculate likelihood ratios accounting for correlations across loci as well within loci. This procedure will outperform even the procedure discussed in this section.

## IX. SAMPLING, REFERENCE POPULATIONS, AND SUBPOPULATIONS

In Section II, I indicated that 5% of the samples in a laboratory's database have type AB blood. In Section VI, I calculated a match proportion of 11.1% (for one of the bands at D17S79). I pretended that these proportions were probabilities. They are not. They are frequencies in the database. As such, they are subject to statistical sampling variability. It is possible to handle this variability (see Berry 1991), but laboratories do not. Some take what they perceive to be a conservative approach (in the sense that it leads to a larger match probability than would a correct approach) by using an upper confidence level of the match probability instead of the sample match proportion. This may indeed be conservative, but to my knowledge no one has demonstrated it.

To calculate a match proportion, laboratories need a reference population. The standard is to use the race of the suspect (or, as in Castro, the race of the victim). This makes no sense. A match proportion is calculated assuming the suspect is *innocent*. So the appropriate reference is the race of the criminal, assuming the criminal is *not* the suspect. In the (unusual?) case that the defense and prosecution agree on the reference population, that population could be used. If they do not agree, the individual populations should be pooled. Perhaps better, a calculation could be made for each possible population, and the jury can choose whichever they like or weigh them appropriately.

A related issue that is regarded by some defense counsels as important is usually a red herring. Suppose the reference population is not in Hardy-Weinberg equilibrium. One way this can happen is that it is a mixture of subpopulations; each subpopulation may be in equilibrium, but with different gene frequencies. For example, suppose a suspect is Haitian and black. A database is available for blacks but not for Haitians. Is it reasonable to use the black database? The answer depends on what is known about the race of the criminal. *The race of the suspect is irrelevant*: We *know* the suspect's phenotypes! If the criminal is known to be Haitian (assuming the criminal is not the suspect), then the lack of data on this subpopulation is a problem. (There are appropriate ways of addressing this problem but none have been published.) But if the criminal (again, assumed not to be the suspect) is known to be black, there is no problem: The black reference population applies. If the race of the criminal is unknown, then one should average over the various possible reference populations as described above.

## X. SUMMARY AND CONCLUSION

1. Uncertainty should be communicated to a court using a likelihood ratio. The numerator of the likelihood ratio is the probability of the genetic evidence actually observed assuming the suspect and crime samples are from the same individual. The denominator is the corresponding probability assuming the samples are from different individuals. Most people confuse a likelihood ratio with the odds that the individuals are identical. Converting a likelihood ratio into odds of identity requires Bayes' rule, which in turn requires assessing a "prior probability" of identity.
2. The inverse of the matching proportion when using a match/binning procedure is a likelihood ratio, but the definition of match that is applied to a suspect's band weights must also be applied to band weights in the database. Otherwise the results are not interpretable.
3. Correctly applied, match/binning is not a horrible procedure, but it has basic flaws. These flaws can serve to exonerate guilty individuals. Match/binning sometimes inflates and sometimes deflates the strength of DNA evidence.
4. In calculating probabilities, one should be guided by the observed distributions of measurement errors. For RFLP analysis, these are bell-shaped. The resulting likelihood ratio is considered in Sections VII and VIII. This calculation accommodates band shifting in a formal way.
5. The issue of subpopulations is usually a red herring.
6. PCR technology is not subject to the same considerations of measurement error as RFLPs. However, sequences identified using PCR are subject to the same

statistical considerations as are other discrete systems such as red cell antigens, and as discussed in Sections II–V.

By way of conclusion, I will answer a question posed by the Editor as to whether I would testify for defense or prosecution in a case involving DNA profiling. I will assume the method of inference used by the laboratory is match/binning. First, however, I want to point out that an issue cited in (3) above is not addressed by this question: There are cases that never reach trial because match/binning has given equivocal results (by virtue of its arbitrary definition of "match"). I do not know the incidence of such cases, but I suspect it is much greater than we are led to believe by the FBI and others.

I could never defend match/binning in court or otherwise as being good statistics. In most criminal cases, the calculations I described in Sections VII and VIII would support the match/binning conclusion (at least qualitatively), although the likelihood ratio could be larger or smaller. Given this result, I could then testify for the prosecution. There may be a small fraction of cases in which the match/binning conclusion is wholly out of line with a correct approach and I could then testify for the defense. (In practice, I have always declined to testify for both prosecutions and defenses.)

## REFERENCES

Balazs, I., M. Baird, M. Clyne, and E. Meade. 1989. Human population genetic studies of five hypervariable DNA loci. *Am. J. Hum. Genet.* **44**: 182.

Berry, D.A. 1990. DNA fingerprinting: Can it prove guilt? *Chance* **3**: 5.

———. 1991. Inferences using DNA profiling in forensic identification and paternity cases. *Stat. Sci.* **6**: 175.

Berry, D.A., I.W. Evett, and R. Pinchin. 1992. Statistical inferences in crime investigations using deoxyribonucleic acid profiling. *J. R. Stat. Soc. Ser. C.* (in press).

Evett, I.W., J.K. Scranage, and R. Pinchin. 1991. An efficient statistical procedure for interpreting DNA single locus profiling data in crime cases. *J. Forensic Sci. Soc.* (in press).

Grunbaum, B.W., S. Selvin, N. Pace, and B.A. Black. 1978. Frequency distribution and discrimination probability of twelve protein genetic variants in human blood as functions of race, sex, and age. *J. Forensic Sci.* **23**: 577.

Lander, E.S. 1989. DNA fingerprinting on trial. *Nature* **339**: 501.

Office of Technology Assessment, U.S. Congress. 1990. *Genetic witness: Forensic uses of DNA tests.* OTA no. BA-438. U.S. Government Printing Office, Washington, D.C.

Weir, B.S. 1991. Independence of VNTR alleles defined as fixed bins. Unpublished report of Program in Statistical Genetics, Department of Statistics, North Carolina State University, Raleigh, North Carolina.

# Public Policy for Forensic DNA Analysis: The Model of New York State

JEROO S. KOTVAL

New York State Legislative Commission on Science and Technology
Albany, New York 12248

## INTRODUCTION

Since the introduction of the first DNA-based forensic tests in the United States in the late 1980s, the courts, the criminal justice system, and policy makers have become increasingly aware of the unique complexity of issues surrounding this new technology. New York State has been a leader in evaluating the benefits and concerns raised by DNA-based forensic tests and in attempting to implement well-thought-out public policy. The deliberations and policy conclusions reached in New York State should serve as a model for other states. However, the introduction of the FBI as the major participant in this area has dramatically altered the political equation. The central policy question is no longer whether licensing and oversight of laboratories that perform this test should be required, but rather who should control the regulation, licensing, and oversight of these laboratories. Serious privacy concerns stemming from the possession of DNA by the criminal justice system—especially the FBI—have received inadequate exposure and consideration.

This paper provides a historical perspective on the public policy questions raised in New York State to illuminate how those questions were shaped and how they ought to be posed unfettered by concerns of "territoriality" and political advantage. A historical perspective also shows how readily the discourse can move from considerations of public policy to those of politics. There is a lesson here about the necessity for vigilance among an informed and alert citizenry. Finally, outlined in this paper are aspects of a sound public policy response to the use of DNA-based forensic tests by the criminal justice system.

## HISTORICAL OVERVIEW OF THE PUBLIC POLICY DISCOURSE ON DNA-BASED FORENSIC TESTS IN NEW YORK STATE

The New York State Assembly held public hearings in October 1988 to investigate the application of "DNA fingerprinting," at which there was near-universal praise for the investigative powers of this technology. Except for comments from a few

---

The views expressed in this paper are solely the views of the author and do not represent the views of any member of the New York State Legislature or of any individual on the staff of the New York State Legislature.

*DNA on Trial: Genetic Identification and Criminal Justice*

defense attorneys about quality control and quality assurance of the test results, the testimony was unalloyed by concerns about privacy emanating from possession by the criminal justice system of an individual's hereditary material. Indeed, the banking of blood samples from which DNA could be extracted at any future time was recommended to the legislature as a valuable tool for law enforcement.

A major report was issued in September 1989 by a panel of experts convened by the governor's Director of Criminal Justice Services (Report of the New York State Forensic DNA Analysis Panel 1989). The report provided a long list of proposals for implementing a forensic DNA technology program in New York. These recommendations focused on three principal issues:

1. *The characteristics of a model accreditation program for forensic laboratories performing DNA analyses.* The report considered specific minimum criteria for laboratory accreditation. Established, for example, were (a) standards for sample handling and extraction of DNA from the sample; (b) standards and procedures for analysis, interpretation, and coding of data; (c) minimum required scientific controls and record keeping; and (d) minimum qualifications and training of the laboratory personnel.
2. *The formation of a Scientific Review Board that would certify new DNA-based technologies as and when they might be developed and before they were introduced into the courts.* This board would also serve as a repository of expert advice to the courts, if such advice was requested. The Scientific Review Board would be composed of scientists with expertise in the disciplines of population biology, molecular biology, and the forensic sciences; it was recommended that the chairperson be a molecular biologist with special competence in forensic applications of the technology.
3. *The creation of an Advisory Committee that would set guidelines and procedures to prevent abuse of this powerful technology, as, for example, in the implementation of data-banking of test results.*

The governor's panel report carried weight because of competent representation from the various constituencies with expertise involved in the use of forensic DNA technology. Representatives included research and forensic scientists, prosecuting and defense attorneys, policy makers, legal scholars, and law enforcement personnel. This report also gave first recognition to the fact that DNA contains the entire hereditary information of the individual and that its use would pose problems of invasion of privacy if adequate protections were not put into place.

Another assembly hearing was held in November 1989, at which special attention was paid to the lessons learned from a Frye hearing in a case adjudicated earlier in that year: *People of the State of New York v. Castro* (545 N.Y. 2d 985). The defendant was charged with murder, and the forensic DNA test results from a private laboratory declared a match between the blood found on his person and that of the victim. Some of the attorneys involved in this case had been to the Banbury Conference (Long Island, New York) held in the fall of 1988, where for the first time a gathering of scientists and other scholars had raised technical, ethical, social, and constitutional concerns about this new technology. Seasoned by discussions at the Banbury Conference, these attorneys marshaled a host of expert witnesses to challenge successfully the forensic DNA test on the basis of its having been unreliably performed and hence not introducible as evidence. The discrediting of test results that were a major piece of evidence in a murder trial, together with the governor's panel report, alerted lawmakers to the need for accreditation and validation procedures and to the harm that could be caused by hasty use of this test in the courts. It also made clear that without adequate population data and well-thought-out controls, these tests could not be considered reliable evidence in criminal cases. Attention was also drawn to the fact that some of these laboratories were holding the test reagents as trade secrets, thereby precluding meaningful scrutiny of their test results by the courts.

Some months after the second assembly hearing, in April 1990, the New York State Legislative Commission on Science and Technology (1990) issued a report with a particular focus on the ethical and social considerations raised by DNA-based forensic tests. This report pointed out the advances likely to be made by the Human Genome Project and the potential concerns about genetic privacy raised by this program. It suggested that new technologies might make personal genetic information more widely available.

A well-crafted bill was introduced soon thereafter in the New York State Legislature with the intent of establishing a forensic DNA-testing program. The bill required that all laboratories submitting their tests to New York State courts be accredited by the New York State Department of Health. The proposal was not submitted to the governor for his consideration because of the FBI's opposition, although the measure passed both houses of the legislature unanimously.

The FBI is a leader in developing DNA-based forensic tests and is rapidly moving to create a database containing test results that can be accessed by state crime laboratories interested in interstate tracking of suspects. The choice about an accreditation and validation program was presented to key New York State legislators by the FBI: If the state were to require the FBI test to be validated and licensed by state authorities before introduction into New York State courts, the FBI would refuse both to do the test for New York State and to allow it access to the nationwide FBI data bank.

A revised bill was introduced, eliminating the need for the FBI to be accredited and containing a requirement that the regulation of forensic DNA laboratories be shared jointly by the New York State Department of Health and the Division of Criminal Justice Services. However, the proposal was not reported from the Republican-controlled senate due to intense opposition from the criminal justice community in New York. Well-thought-out public policy is currently mired in battles over territorial control.[1] The basis of conflict appears to be a reluctance of all segments of the criminal justice system to allow any authority for regulation and/or quality control to exist outside the criminal justice agencies, i.e., in the state health department.

## CONSIDERATIONS FOR PUBLIC POLICY

The preceding legislative history of public policy for DNA-based forensic tests should indicate that the thoughtful recommendations of the governor's panel, and the ethical and civil liberties concerns raised in the report issued by the Legislative Commission on Science and Technology, consistently point to the following crucial requirements for a sound forensic DNA program in any state:

1. A thorough system of validation and accreditation, and a review board composed of scientists to determine the validity of each new test and set the minimal guidelines necessary for accreditation
2. A committee with accountability to the legislature and the governor, with responsibility for reviewing the ethical and constitutional concerns arising from the use of forensic DNA-based tests
3. National legislation to require that biological samples collected for investigative purposes be destroyed after a fixed period of time

The rationale behind each of these recommendations is elaborated below.

## ACCREDITATION AND VALIDATION

### Need for a System of Validation and Accreditation

Admission of DNA-based tests into the courts has been on a case-by-case basis, frequently requiring lengthy Frye hearings. In an age of rapidly changing DNA technology, this situation presents concerns about (1) an absence of uniformity in the strictness of admission standards since the rigor with which different courts may question the admissibility of a test would depend on the admixture of lawyers and expert testimony that the parties to the litigation could marshal; (2) the cost of Frye hearings, which would be required, since each new test could be open to question; and (3) the fact that Frye hearings speak only to the validity of the testing procedures, based on their general acceptance by the cognizant scientific community. Frye hearings do not guarantee that an individual testing procedure has been performed satisfactorily and interpreted competently. This assurance is especially important, since test results often claim that identity can be certified with odds such as one in a million.

---

[1]*Editor's note:* The governor of New York State has recently altered the political landscape by indicating his support for the use of DNA identifications in various criminal investigations.

To date, almost none of the forensic laboratories in the United States are required to be accredited or licensed before they introduce their tests in courts. Furthermore, few laboratories are supervised by individuals with training or experience in population biology or the sophisticated techniques of molecular biology used in DNA-based forensic tests.

Although an accreditation program for forensic laboratories is available through the Association of Crime Laboratory Directors, participation is on an entirely voluntary basis. The FBI has also moved to adopt and promote quality assurance guidelines via a policy of requiring state laboratories to submit to these guidelines before they can participate in the data-banking network planned by the FBI. This enforces participation in an FBI-controlled quality assurance program, which by the agency's own admission represents "the minimum quality assurance requirements for DNA" (Technical Working Group on DNA Analysis Methods and California Association of Criminalists Ad Hoc Committee on DNA Quality Assurance 1991). It does not preclude a state from requiring its own, perhaps more stringent, oversight of forensic laboratories submitting tests to the state's courts. This oversight role should correctly be assumed by states, since criminal justice issues are constitutionally under state jurisdiction. However, as yet no state requires that a laboratory be accredited before it can introduce its test results in the courts.

Experience with proficiency testing of medical diagnostic laboratories in New York State indicates the urgent need for a similar program for assuring quality control of DNA-based forensic tests through the establishment of standards and a competent external evaluation regime for the certification of laboratories. As an example, proficiency testing of cytopathology laboratories conducted by the New York State Proficiency Testing Program indicated that at the beginning of the testing program in 1971–1972, 47% of the 232 laboratories failed to measure up to the standards of the test. Over the years, the pass rate has improved considerably, very likely due both to a greater familiarity with the testing process and to the remedial education required prior to retesting. However, even as late as the 1982–1984 testing period, 20% of the 209 laboratories tested failed to pass a proficiency test on the first attempt (Collins and Patacsil 1986).

The types of errors responsible for most failures on proficiency tests are said to be: (1) equipment or reagent errors or failures, (2) clerical errors such as those involved in copying data, and (3) personnel errors such as those involved in carryover and contamination of separate experimental regimes, errors from not following protocols diligently, or errors in interpretative reading (L. Flaherty, New York State Department of Health, pers. comm.).

Even though errors in chain of custody are less likely with forensic samples than with medical samples that arrive in batches at the diagnostic laboratories, the likelihood of human errors must be assumed to be significant compared to the very large odds claimed for certainty of an identifying match. Formal recordkeeping to discover the frequency of such errors should be required and made available to all review boards charged with oversight of forensic DNA-based tests.

Population biology issues, especially important in interpreting the meaning of a match on the test, are unsettled, and the controversy surrounding subsetting by racial and ethnic background should be of particular concern to local criminal justice jurisdictions (Chakraborty and Kidd 1991; Lewontin and Hartl 1991).

## Essential Features of a Good Accreditation and Validation Program

At a minimum such a program should include:

1. *A review board (such as the Scientific Review Board) comprising population biologists, molecular biologists, and forensic scientists (the last with expertise in molecular biology) to determine whether all new DNA-based tests used in the state are sound.* Such a board should have power to review all materials used by a forensic DNA-testing laboratory, precluding private laboratories from holding their reagents as trade secrets. This board should also determine the scientific controls required for each test and promulgate minimal standards for the operation of forensic laboratories. Such a board should also be available to provide expert testimony to courts upon request.
2. *A proficiency testing and licensing program for all government and private laboratories wishing to submit their tests in the state's courts.* Such a testing program should be conducted by scientists with expertise in relevant disciplines and with experience in administering proficiency tests. For most states, such expertise is likely to be available in their health departments, which traditionally are charged with proficiency testing and accreditation of clinical laboratories. It is also possible that the forensic sciences expertise of the criminal justice system could be profitably harnessed to jointly regulate and monitor the functioning of forensic DNA laboratories. The involvement of scientific expertise from outside the criminal justice system is especially important to guard against the possibility of a conflict of interest and to ease public concerns about openness in the testing and licensing of forensic DNA laboratories.

Final authority for certifying that all test procedures have been satisfactorily performed and accurately interpreted should reside with a scientist having Ph.D. training in molecular biology and expertise in forensic sciences. The present dearth of such individuals does not obviate the need for reassurance in the interpretation of a complex test such as those being discussed. Such a uniquely trained laboratory chief could be shared with other laboratories on a rotating basis.

The aura of territorial competitiveness that pervades current public policy discourse about validation and accreditation programs tends to cloud the necessity for such programs. The intent of these programs should be not merely to expose unreliable execution and interpretation of DNA-based forensic tests and penalize those who fail to perform adequately, but also to identify weaknesses in a forensic DNA program and rectify these as far as possible through feedback and remedial education. Laboratories should be penalized or closed down only as a last recourse.

## ETHICAL AND CIVIL LIBERTIES CONCERNS

### Concerns about DNA-based Identification Methods Becoming a Routine Identification Tool

Today, DNA-based identification tests are permitted for their usefulness in implicating or exonerating suspects involved in some of society's more heinous crimes. It seems highly likely that as the tests become available as inexpensive automated kits, their uses will expand.

Consider the history of fingerprinting. Originally, it was introduced to provide valuable data in criminal investigations, but it has become a routine procedure in every arrest. Right now, even as we are told that forensic DNA analyses are to be used only for identifying violent offenders and rapists, an Office of Technology Assessment report finds that many state and local crime laboratories are considering applying DNA tests for hit-and-run and robbery cases if crime scene evidence is available (Office of Technology Assessment 1990). There has been little debate about whether this is good social policy.

### Concerns about Genetic Privacy

The ideal DNA-based identification test is not yet at hand. The testing procedures are changing rapidly, and with each new test, the public's concerns about privacy issues and the fair use of the testing procedures need to be addressed. This is of special concern in the day and age of the Human Genome Project, which is conservatively estimated to be a $3 billion enterprise and which has as one of its goals the complete sequence of human DNA.

What are the implications of the Human Genome Project to the availability of DNA to the criminal justice system, as, for example, to federal and state investigative bodies? There have been reports that the mere knowledge of an inherited medical disability has led to discrimination of individuals by insurers and employers, even before there was any manifestation of the disease (Billings et al.

1992). It is sobering to consider what could be done in time with stored DNA samples, as the Human Genome Project identifies an increasing number of sequences involved in determining neuropsychological and personality traits and other heritable tendencies, and as the tests become cheaper. The track record of the FBI in the area of respect for civil liberties and the public's unease about the possible misuse of stored DNA in the possession of an investigative agency are legitimate concerns of public policy.

It would seem wise to consider national legislation to guarantee the destruction of the DNA sample taken from an individual within a fixed time after the completion of the investigation. Furthermore, legislation should specify that no test other than the one for forensic identification shall be performed on the DNA sample taken for forensic use. Medical diagnostic tests should not be performed on DNA taken for forensic use unless a court order or the consent of the individual has been obtained.[2]

## Concerns about the Misapplication of DNA-based Identification Tests

There are limits to what legislation can do, and there is a serious need for education of the legal community about the limitations of DNA-based forensic tests. The tendency to become overly impressed by the power of forensic DNA analysis is a concern in individual prosecutions. *People of the State of New York v. Wesley* (1992; ___, A.D. 2d. ___.) illustrates this point. The defendant, a mildly retarded man, had been convicted of murder because a match had been declared between the DNA in blood found on the defendant's person and that of the victim. The defendant claimed that another person was present who actually committed the crime, and indeed a bloody sneaker print at the crime scene did not match the footwear on the defendant. Even so, this man was convicted of murder. The appeal has been submitted but has not yet been heard. This is a good example of how more can be read into the test than is warranted. A forensic DNA test does not convict a person. At best, the test reveals identity. In this matter, all that can be said is that although the DNA test may indicate that the defendant was present at the crime scene, it cannot specify him to be the murderer.

Cases such as this point to the urgency for education of the legal community about the meaning of forensic DNA-based test results. It is a serious ethical concern that a match can be concluded to mean that the suspect is guilty.

---

[2]*Editor's note:* There is general consensus that retention of DNA samples by governmental agencies may pose important civil liberty and privacy issues, which might preclude this practice. Similarly, the testing of a forensic sample for medical or *any other purpose* may not be acceptable even with strict regulation and informed consent.

## SUMMARY

If wisely used, DNA-based identification methods can be a valuable tool for the criminal justice system not only in identifying the perpetrator of a crime, but also in exonerating the innocent accused. However, the test is complex, the technology is changing rapidly, and the procedure uses the hereditary material of all individuals. Therefore, a sound public policy response to the forensic use of DNA-based identification methods must include:

1. Vigilant proficiency testing and accreditation of forensic laboratories by scientists with expertise in proficiency testing
2. Evaluation, discussion, and debate about each new methodology as it becomes available (This should include exchanges not only about its validity and usefulness, but also about the ethical and social implications of its widespread adoption, again with special emphasis on education of the legal community.)
3. National legislation to forbid the banking of DNA samples taken for investigative purposes and/or the use of these samples for any further tests

## ACKNOWLEDGMENTS

The author expresses particular thanks to Jan Fink, Esq., Patricia Murphy, Ph.D., and Barry Duceman, Ph.D. for helpful discussions during the preparation of this paper.

## REFERENCES

Billings, P., M. Kohn, M. de Cuevas, J. Beckwith, J. Alper, and M. Natowicz. 1992. Discrimination as a consequence of genetic testing. *Am. J. Hum. Genet.* **50:** 476.

Chakraborty, R. and K.K. Kidd. 1991. The utility of DNA typing in forensic work. *Science* **254:** 1735.

Collins, D.N. and D.P. Patacsil, Jr. 1986. Proficiency testing in cytology in New York: Analysis of a 14 year state program. *Acta Cytol.* **30:** 633.

Lewontin, R.C. and D.L. Hartl. 1991. Population genetics in forensic DNA typing. *Science* **254:** 1745.

New York State Legislative Commission on Science and Technology. 1990. *Forensic DNA analysis: Scientific, ethical and social considerations.* Albany, New York.

Office of Technology Assessment. 1990. *Genetic witness: Forensic uses of DNA tests.* U.S. Government Printing Office, Washington, D.C.

Report of the New York State Forensic DNA Analysis Panel. 1989. *DNA.* Albany, New York.

Technical Working Group on DNA Analysis Methods (TWGDAM) and California Association of Criminalists Ad Hoc Committee on DNA Quality Assurance. 1991. Guidelines for a quality assurance program for DNA analysis. *Crime Lab. Digest* **18:** 44.

# The Impact of DNA-based Identification Systems on Civil Liberties

PHILIP L. BEREANO

University of Washington
Department of Technical Communication
Seattle, Washington 98195

## THE NATURE OF DNA-BASED IDENTIFICATION SYSTEMS

Several different contexts distinguish the use of DNA-based technology for identification, including (1) the *civil* context, such as the use of typing to establish paternity; (2) *individualized criminal prosecution*, where there is an attempt to match up a tissue sample obtained at the scene of a crime with DNA from an individual who is the accused in a prosecution; and (3) the establishment of a *bank or library* that will have DNA profiles on record and that could be used for matching individuals in the future, for investigative purposes. These three situations present increasing concern from a civil liberties point of view. A civil case presents fewer concerns than a criminal case, because the power of the police and the state are not behind the DNA investigation. The problems are most acute in the third situation, the DNA data bank. Unlike the implications in much of the material put out by the FBI and other proponents (see OTA 1990), the use of DNA typing in individual prosecutions *need not* lead to the establishment of a DNA identification library with its greater threats to our liberties.

## THE CONTEXT: AN INCREASINGLY TECHNOLOGICAL SOCIETY

Technologies are not value-neutral; they usually embody the perspectives, purposes, and political objectives of powerful social groups. Although the traditional ideology in this society proclaims that science and technology are value-neutral and objective, during the past two decades an increasing number of social commentators have helped to dispel this myth (Bereano 1976). Under this traditional view, the only problems caused by technologies are either "externalities" (unintended side effects) or abuses. However, because technologies are the result of human interventions into the otherwise natural progression of activities, they are themselves actually *imbued with intentions and purposes*. Current technologies do not equally benefit all segments of society (and indeed are not intended to do so), although to maximize public support for these developments and to minimize potential opposition, their proponents rarely acknowledge these distributional ramifications.

The United States is a society in which the differential access to wealth and power has been exacerbated during recent years. Because technologies are intentional interventions into the environment, those people with more power can determine the kinds of technological developments that are researched and implemented. Because of their size, scale, and requirements for capital investments and for knowledge, modern technologies are powerful interventions into the natural order. They tend to be the mechanisms by which already powerful groups extend, manifest, and further consolidate their powers. Thus, technologies themselves are not neutral; they are social and political phenomena. Recombinant DNA technologies, along with other technologies such as computerization (see Schellenberg 1990), exhibit these characteristics.

The dominance of science and technology in the ideology of this society rests on three promises which have their root in a mythic claim that science and technology are *omnipotent*. The first two of these promises are publicly discussed, and the third is something that we acknowledge in our darker moments. They are (1) the claim of *infallibility*, that technologies will help us reach perfection; (2) the *"technological fix,"* that technologies can be derived which will help us solve social problems that are too sticky for us to deal with politically or socially; and (3) the idea that technology offers means of *social control*, particularly for those in power.

The notion of technological infallibility persists despite widely publicized disasters that we have all been witness to in recent years, i.e., the Challenger explosion, Chernobyl, Bhopal, and so forth. This mythology is perpetuated by making what I think is a false distinction between technical and human fallibility. The failures are always blamed on the humans; the machine is perfect. However, the true nature of science and technology is that they only exist in terms of humans and human institutions. To say otherwise is to anthropomorphize technology, making it into some kind of a being, as if it had a life of its own—and it does not. It has a life that is determined socially by political groups in the government, in the corporate sector, and perhaps, in the citizenry.

The notion of the technological fix is exacerbated by the "gee whiz" reporting of the media. The proponents of new technologies often find this approach consistent with maintaining their power and their own ideology. It pacifies the populace who legitimately do want solutions to the problems of crime, drugs, social insecurity, and an inadequate health care system by saying that the engineers will come up with a magic to make everything better. It deflects attention from the social roots of problems, which might suggest that redistributive policies, making power and wealth more available and equal, need to be considered.

The FBI materials produced on the subject of DNA-based identification banks exhibit the technological fix approach (Hicks 1989a,b). There is no recognition that violent crime may be correlated to social factors in a way that the existence of new technologies will *not* serve as a deterrent. The argument that DNA-based identification will deter sex offenses is purely wishful thinking, not based on any supporting evidence. All the elements of the course put together by the FBI, most of its published articles, and much of the testimony of its officials ignore relevant social factors (not just civil liberties), as if the technology existed in isolation (Hicks 1989a,b). The use of DNA identification is hence being grossly oversold; for instance, the identity of the perpetrator is *not* an issue in some violent crime (e.g., "date rape"), and much crime leaves no identifying tissues behind (e.g., burglary).

In regard to social control, we must understand that proposals for DNA-typing systems are being put forward at a time of an unprecedented testing hysteria. The daily newspapers illustrate the development of an ominous pattern in our society in which people are increasingly calling for all sorts of scientific "testing" of others, such as employees, university athletes, members of the armed forces, prostitutes, or sex offenders. Once a technological program like DNA identification is established for a pariah group such as sex offenders, inevitably there will be pressures to extend it to yet other groups and also to allow access to the information by increasing numbers of individuals and institutions who claim that they have a "need" for the information contained therein.[1]

This has already happened in the state of Washington; although the nation's first legislation to establish a DNA data bank, by King County (Seattle), was predicated on the high recidivism rate among sex offenders, legislation adopted by the state legislature would also include mandatory testing of those convicted of "violent crimes" (R.C.W. 43.43.752), many of which (like homicides) have the lowest rates of recidivism or leave no tissues behind. There have been reports that the British were planning to do DNA testing on IRA "terrorists" (R. Hubbard, pers. comm.), and we know how the label of "terrorist" (or "bandit," "deviant," or "gang member") is affixed to dissident groups as part of a control strategy by those in power.

We can also surely expect attempts at forced testing of individuals who are not criminals, for their own good or the good of society, and the insurance industry, corporate employers, politicians, and government bureaucrats will be in the lead pressing for such programs. When the technology becomes cheap enough (by having expensive programs like police work absorb a lot of front-end costs), proposals will be made to do a DNA "print" from every newborn baby's "heel-stick" blood sample, just in case they should ever become amnesiac, lost, or abducted. In other words, the technology of DNA typing can easily lead to increased social control by powerful elites over *potentially* "unruly" citizens.

These scenarios are not hysterical but are based on strong historical analogy where other control technologies have inevitably been extended (see discussion of Social Security number in Hoeffel 1990, p. 535). If it were to happen, it would not be as part of a crude and obvious totalitarianism, but as an "efficient" and "rational"

---

[1]*Editor's note:* These points are additionally illustrated by the variation among states in rules for which kinds of criminals, convicts, and suspects will be compelled to give samples to be used for DNA identification. In addition, the Armed Services-based DNA-banking program has already been solicited by the National Cancer Institute to cooperate in genetic epidemiology "research" (E. Juengst, pers. comm.).

means of reaching socially desirable ends (like dealing with violent crime, drug abuse, missing people) (see Hicks quoted in Hoeffel 1990, p. 537).

Civil liberties depend essentially on a celebration of notions of diversity. Technological rationality, on the other hand, depends on notions of uniformity: being able to put things in a small number of categories, keeping them there, and tracking and monitoring them. Efficiency has always been the value that the bureaucrats most prize. Despite the importance our culture places on the values of efficiency and rationality, we also honor conflicting values: individuality and diversity, civil libertarian concepts. We must be ready to recognize that freedom often flourishes best in an inefficient society, that many of the freedoms that we enjoy as a practical matter are exercised in the interstices of the kind of matrix that governments, with their new technologies, are able to establish to control and monitor their citizens.

Living with conflicting values is not often well understood in our society, but it is built in to the fundamentals of our system. For example, our criminal system does not have the identification and conviction of the guilty as its only goals. If it did, we would not have developed the rules of evidence that we have, such as preventing spouses from testifying against each other. A person can invoke the privilege that prevents spouses from testifying against spouses, although presumably a lot of spouses' testimonies would actually be helpful in convicting people who are guilty. In this case, other values—a belief that the privilege fosters family harmony—bar efficient prosecution. We have the "exclusionary rule" that forbids the admittance of evidence which the police have illegally seized. Much of this evidence would certainly help to convict the guilty, but an important competing value is that we do not want the police to break the law. Our system is one of balancing in these situations.

Many of these same considerations apply to DNA-based identification. Of course, such a system would help find guilty people, but if the system is going to compromise other important social values, perhaps we should not implement it.

## CIVIL LIBERTY PROBLEMS INHERENT IN DNA IDENTIFICATION SYSTEMS

For purposes of discussing these important questions in a more structured fashion, we can organize the civil liberty and social concerns into two main categories, "genetic privacy" and "genetic redlining," both relating to a perceived threat to the American notion of the proper autonomy of the individual. The concepts of individual autonomy and privacy, reflected from the earliest days of American society in the Fourth and Fifth Amendments, have been additionally elaborated within the past decade or two in a variety of court cases relating to the Ninth Amendment, most notably by the Supreme Court beginning with *Griswold* v. *Connecticut* (381 U.S. 479, 1965).

Congress approved the Federal Privacy Act in December of 1974. As originally drafted, the Act's principles would have covered the information systems of private industry and of state and local governments, as well as those of the federal government. As approved, the law only applies to federal agencies and the private organizations it does business with. A second major reduction in the scope of the act involves the elimination of federal law enforcement and intelligence agencies from many of its requirements.

The Privacy Act would have had a significant impact if it had been faithfully implemented by Presidents Nixon, Ford, Carter, Reagan, and Bush. One provision, for example, requires that federal agencies collect only information that is relevant and necessary, but this restriction has not been heeded by the military, the FBI, the Secret Service, the Department of Health and Human Services, and other agencies. A second provision requires that, to the maximum extent feasible, federal agencies obtain information directly from the individual they are interested in, but the linking of databanks to exchange information runs afoul of the spirit of this provision. A third item requires federal agencies to inform the individual why the information they are seeking is required, how the information will be used, under what statutory authority it is being solicited, and what penalties will be imposed if the individual declines to respond. These strictures are not always followed. All individuals have the basic right to inspect and correct their federal records, but not all people know this (or even that a record on them may exist) (Privacy Act of 1974, 5 U.S.C. 552a).

The Act also limited compulsory disclosure of a Social Security number to a local, state, or federal agency as a condition of receiving services or benefits unless such use is authorized by law. Congress has repeatedly allowed the use of the Social Security number (e.g., as a tax identification number). Other currently authorized uses requiring giving the number include obtaining a driver's license, installing a telephone, joining the Chamber of Commerce, giving blood, and obtaining a library card.

To understand the ways in which DNA typing may threaten privacy and lead to increasing social control, stigmatization, and discrimination, we have to avoid analyzing the systems in isolation and appreciate that they are part of a much larger set of events, such as the project to "map" or "sequence" the human genome (the most costly single federal biological research program) and the "testing mania" noted above. The Congressional Office of Technology Assessment report on the genome project (OTA 1988) recognizes that "genetic information has often been used for political purposes in the past; information arising in genome projects could be similarly misused." As the genome project unfolds, we can expect its information to lead to new legal categories of neglect by parents, negligence claims against carriers of "abnormal" genes, malpractice charges against medical personnel and genetic counselors, and similar developments. The current debate surrounding calls to test all health care providers for HIV infection illustrates how public health and civil liberties may be compromised for little, if any, real benefit. (The data do not, in any way, support such a program [Altman 1991; Hilts 1991].) Rational and efficient arguments will be made to extend DNA libraries and use them to combat new problems that will be defined as "genetic" (e.g., alcoholism, schizophrenia, criminality, and homosexuality).

FBI initiatives, prodding enthusiastic state and local law enforcement officials, have led at least 11 states to enact requirements that DNA typing be performed on convicted offenders (as of January 1990): Arizona, California, Colorado, Florida, Illinois, Iowa, Minnesota, Nevada, South Dakota, Virginia, and Washington. A number of additional states have been considering data banking: Connecticut, Massachusetts, Michigan, Indiana, and Ohio.

**Genetic Privacy**

Once the capability to perform testing and to collect, store, and correlate information exists, we can be sure that pressures (from both government and segments of the private sector) will mount to do so. There will be calls to sacrifice individual privacy and confidentiality for the greater social good, to require involuntary submission of blood and tissue sampling, and to reduce control by individuals over their personal information once it is placed in the system and is coursing through interlinked computer networks.

In the autumn of 1990 (and again in 1991), Representative Conyers introduced a bill (originally H.R. 5612) labeled the "Human Genome Privacy Act" that would have provided some protection for genetic information collected by government agencies (regarding access, disclosure, accuracy), analogous to legislation protecting credit information. It was inadequate in protecting privacy because it did not restrict the *collection* of such information, which ought to require substantial justification before bureaucracies are allowed to engage in such activities. This is because privacy is invaded, initially, by the forced taking of the blood or tissue sample to obtain the information. Additional privacy concerns arise, for example, when we realize that an individual's blood contains literally thousands of bits of data, e.g., an individual's heredity, genetic history, disease resistance, and drug use. There is strong concern that having this kind and amount of information about an individual turned over to the government and kept on file presents tremendous opportunities for very substantial civil liberty abuses.

The bill also failed to cover the private sector; we can be sure that insurance companies, employers, and other powerful private entities will make strenuous efforts to obtain such information if it exists in government agency files. Recent litigation against the largest purveyor of consumer credit information, TRW, although now settled, shows that current information data banking is not protecting individual privacy.

The government agencies have claimed they would only focus on the parts of DNA that they could use to identify people, and they imply that they would restrict access by private entities such as insurance companies (Hicks 1989a). Frankly, however, why should we be willing to accept the word of public officials on such matters? During the past 20 years, we have uncovered so many instances of government officials saying one thing and doing something else (e.g., the Gulf of Tonkin, Cambodia, and Nicaragua incidents) that we should demand adequate institutional checks to assure that these protections are handled adequately.

Originally, proponents only talked about using DNA-based identification technologies in situations that would garner public sympathy—to track down serial murderers, find missing children, reunite broken families, control fiendish sex offenders. In reality, the net of mandatory sampling and testing is inevitably cast wider and wider, to include situations that may not be so appealing to everyone. After all, serial murders are *not* the kind of crime that objectively endangers many Americans, but tying DNA techniques to that scenario certainly gets good press. (Also resulting in good press are the uses of DNA typing to reunite families "disappeared" in Argentina and to exonerate accused criminals, but neither of these beneficial uses requires the establishment of DNA identification libraries under police control.)

**Genetic Evidence**

Evidence has already begun to be assembled of existing cases of discrimination based on genetic screening (Billings et al. 1992). Defining genetic discrimination as "discrimination...solely because of real or perceived differences from the 'normal' genome," discrimination was found in employment (in both public and private sectors), access to social services (again in both public and private sectors), insurability (including life, disability, health, and auto), and health care. Whole families are being stigmatized, including their ancestors and their unborn, with the development of significant subsequent interfamilial stress and recrimination. Individuals have been immobilized in current jobs, residence, and public or private programs, because any change would result in their inability to obtain insurance; they believe they must stay put to continue coverage under their present policies.

Extensive DNA profiling could lead to pressures for eugenics programs that would raise serious ethical and political concerns. We may feel that American society is far too democratic and rational to allow such programs to come to pass, but this view may be naive at a time when neo-Nazism is flourishing. Eugenics ideas keep being reborn, phoenix-like, most recently in the fascination for sociobiology. For example, about 20 years ago, it was claimed that males who had an extra Y chromosome (XYY males) were more likely to be aggressive and thus to become criminals. Studies were done among prison populations and purported to prove this hypothesis. Fortunately, critical scholars scrutinized this research and demonstrated that it was based on poor data, weak logic, and ideological presuppositions (Pyeritz et al. 1977). The implications of this research could have been profound, however, resulting in ignoring the environmental and social factors, such as poverty and oppression, that might be correlated with criminal conduct; instead, it would have favored medical/technocratic approaches. Similarly, at a time of increased pressure to restrict the reproductive freedoms of women, we need to appreciate that some new genetic knowledge may exacerbate restrictions on them; for example, information about fetal alcohol syndrome has lead to vigilante "pregnancy police," socially embarrassing pregnant women having a drink (Gellene 1991; Kantrowitz 1991).

Scholars at the Hastings Center of Ethics and the Life Sciences point out that it is likely that all human beings have sequences of DNA which are "odd" or which might in some sense make the individual "unfit." Despite this huge range of human variability, most public policy approaches want to sort individuals into two categories, labeled "normal" and "abnormal." Many genetic conditions are, in fact, "contingent" and may never manifest themselves. We should be a little more humble about claiming any abilities to prognosticate human traits and conditions. "Genetic forecasts remain statistical, predictions not predeterminations." The authors suggest that "perhaps we should actively encourage greater public acceptance of variation and vulnerability" (Nolan and Swenson 1988).

Often "normal" is a social, not scientific, construct (see, e.g., the development of genetically engineered human growth hormone as a commodity to be marketed to the shortest 3% of "normal"-sized humans instead of being limited to a medical treatment for pituitary dwarfism [Werth 1991]). Billings notes that many individuals (even at high risk) decline voluntary genetic screening even under auspicious conditions because of the potential for discrimination. Persons informed that they may have a genetic "abnormality" are likely to engage in schemes of secrecy, withholding this information from physicians, government, schools, and employers because of its personal nature and the potential ramifications (discrimination) from its release. Another common response will be lying and evasion in order to get social benefits and entitlements which are key to existence in our society. Feelings of desperation and hopelessness will be frequent because of constant fighting against a label of "disability" even when the disability is nonclinical or will never develop.

**CONCLUSION**

Some intrusions into civil liberties are justifiable under our legal system in certain circumstances, where on balance the state has a compelling interest. We should require the proponents of DNA-based identification data banking to justify the claimed benefits of the technology, before it is implemented. They should be required to demonstrate that its use will, in significant or meaningful ways, result either in convictions or deterrence of sexual offenses and other crimes. Such evidence, if it exists, needs to be made publicly available prior to any decisions to implement this technology; the argument that there is likely to be a deterrent value to this technology is weak, in my opinion, given that increased incarceration, police patrols, and other technologies have not had very much of a demonstrable effect.

How significant an increase in successful convictions justifies intruding on our civil liberties? If we had but one additional conviction per year, would that justify establishing a DNA library and invading our civil liberties? Or, for the proper balance, do we need 10? 15? 20? The answer is not written in any rule book; it depends on social values. Every citizen, and the citizenry collectively, would have to make a judgment that a certain infringement of our civil liberties would be worthwhile only if we could expect "X" additional convictions. I am not going to insist that my own calculus be accepted, but it must be recognized by the public that the answer is not something that the scientists can tell us. It is a political judgment, for the body politic to make. People have to realize that they should not abdicate this judgment to scientists and politicians, who may be motivated by very different values and objectives from their own. More robust public debate of these questions is certainly desirable.

## REFERENCES

Altman, L.K. 1991. U.S. backs off on plan to restrict health workers with AIDS virus. *New York Times*, December 4, p. A1.

Bereano, P.L. 1976. *Technology as a social and political phenomenon.* John Wiley & Sons, New York.

Billings, P.R., M.A. Kohn, M. de Cuevas, J. Beckwith, J.A. Alper, and M.R. Natowicz. 1992. Discrimination as a consequence of genetic testing. *Am. J. Hum. Genet.* **50**: 476.

Gellene, D. 1991. Close up, protecting the unborn: Fetal injury at mom's job is at the center of a growing debate. *Seattle Times*, June 9, p. A3 (originally in *L.A. Times*).

Hicks, J.W. 1989a. Federal Bureau of Investigations, testimony before U.S. House of Representatives, Subcommittee on Civil and Constitutional Rights, Committee on the Judiciary, March 22.

―――. 1989b. Conference summary. In *The International Symposium on the forensic aspects of DNA analysis.* Quantico, Virginia, June 23.

Hilts, P.J. 1991. Experts oppose AIDS tests for doctors. *New York Times*, September 20, p. A11.

Hoeffel, J.C. 1990. The dark side of DNA profiling: Unreliable scientific evidence meets the criminal defendant. *Stanford Law Rev.* **42**: 465.

Kantrowitz, B. 1991. The pregnancy police: From the courts to the restaurants, a boon in bossing around mothers-to-be. *Newsweek*, April 29, p. 52.

Nolan, K. and S. Swenson. 1988. New tools, new dilemmas: Genetic frontiers. *Hastings Center Rep.* **18**: 45.

Office of Technology Assessment (OTA). 1988. *Report Brief: Mapping our genes–genome projects: How big? How fast?* (OTA-BA-373). U.S. Government Printing Office, Washington, D.C.

---. *Genetic witness: Forensic uses of DNA tests.* OTA-BA-438. U.S. Government Printing Office, Washington, D.C.

Pyeritz, R., H. Schreier, C. Madansky, L. Miller, and J. Beckwith. 1977. The XYY male: The making of a myth. In *Biology as a social weapon* (Ann Arbor Science for the People Editorial Collective). Burgess Publishing, Minneapolis.

Schellenberg, K., ed. 1990. *Computers in society*, third edition. Dushkin Publishing Group, Guilford, Connecticut.

Werth, B. 1991. How short is too short? *N.Y. Times Magazine*, June 16.

# Genetics, Race, and Crime: Recurring Seduction to a False Precision

**TROY DUSTER**

Institute for the Study of Social Change
University of California
Berkeley, California 94720

> "It is better for all the world, if instead of waiting to execute degenerate offspring for crime, or to let them starve for their imbecility, society can prevent those who are manifestly unfit from continuing their kind." May 2, 1927, Oliver Wendell Holmes for the majority (8–1) in the Supreme Court Decision, *Buck v. Bell.*
>
> "Some people advocate compulsory sterilization of habitual criminals and mental defectives so that they will not have children to inherit their weaknesses. Would you approve of this?" From a 1937 survey of reader opinions, *Fortune* magazine learned that 63% of their readership responded "Yes" regarding the forced sterilization of criminals. (Reilly 1991, p. 125)
>
> In a few score of reported cases, an XYY fetus has been unexpectedly diagnosed during prenatal genetic studies that were initiated because of advanced maternal age or other reasons....In the United States, an informal survey by one of the leading researchers on XYY males, Dr. Arthur Robinson of Denver, showed that about 50% of parents elected to terminate such pregnancies. (Milunsky 1992, p. 58)
>
> Those who cannot remember the past are condemned to repeat it.
>                                                                      George Santayana

The Greeks have left us many powerful images of the inexorable unfolding of human frailty. None is more compelling than that of Ulysses at the mast who, knowing his own weakness, has himself tied to guard against the magnet of the siren's call. Throughout much of recorded human history, there has been a very magnetic siren against which we must also guard—the palliative notion that poverty, crime, poor health, and even "idleness" or unemployment are conditions best explained by reference to the intrinsic qualities of those who are their victims.

In modern times, including especially the last century, the study of the genetic makeup of humans has been deeply colored by the concern to "fix" those of the least privileged in the society who are deemed to be the source of social trouble. Many of the founders, key researchers, early practitioners, and strong advocates of human genetics (F. Galton, K. Pearson, C. Davenport) had an explicit social policy agenda. They were out to "improve" the human genetic stock by getting the rich and successful to procreate more and, conversely, and more ominously, to get the poor to have fewer offspring. Sometimes this resulted in prevention via involuntary sterilization (Reilly 1991); at times it has taken a genocidal turn, as in the infamous Third Reich.

There is already some evidence to suggest that to be ignorant of that history is to be doomed to repeat it, although not in the crass and unsophisticated way of the 1920s. It is true that Singapore today has a *genetic* policy of encouraging the wealthier to breed more and discouraging the poor from breeding (Gould 1985). Harvard Professor of Psychology, Richard Herrnstein, authored a paper in 1989 in which he suggested that *the genetic stock* of the United States would be improved if the wealthy had more children and the poor had fewer children. Even more striking as a reminder of our history, Herrnstein (1989) went on to suggest that unemployment rates might be explained by genetics. However, these are the fringes. It is not such unreconstructed twentieth-century versions of genetic "pollution at the bottom" that should be the main source of a new social concern. Far more seductive is the contemporary situation in which the banner of science, medicine, and forensic precision flaps over the new genetic technologies. For example, the new technologies of DNA identification (discussed elsewhere in this volume) do enhance probability of identification, but equally significant is the "halo effect" that attends the use of DNA technology. Much as the halo of smart bombs and smart missiles inclines us to an antiseptic view of warfare, DNA identification can provide a false sense precisely because of increased precision: Even if we can match the DNA at the scene of the crime with that of the suspect, the case has not been made that the accused is guilty. As the Bill of Rights would remind us, there is far more to proving guilt than positive identification. Yet, there is likely to be an overwhelming tilt against the accused when such technologies are employed. I turn now to a discussion of the larger social and historical backdrop of the connections between genetics, crime, race, and ethnicity.

Although many are aware of the gross abuses during the early part of the century, most of the current advocates, researchers, and celebrants of the putative link between genetics and crime are either unaware of *the social context of that history*, or too quick to dismiss that history as something that happened among the unenlightened. Both formulations miss the special appeal of genetic and eugenic explanations to the most privileged strata of society, those who lay claim to the legacy of enlightenment. It was the president of Stanford University, respected bankers and politicians, governors, university professors, and other respected professionals who favored the sterilization of the "lower forms" of human life, well into the middle part of this century.

Every era is certain of its facts. The heyday of the eugenics movement was no exception, sure that feeble-mindedness, degeneracy, and criminality were inherited. In 1912, the American Breeder's Association, an organization of farmers and university-based theoreticians, created a "Committee to Study and to Report on the Best Practical Means of Cutting off the Defective Germ Plasma in the American Population." It was a five-man committee, chaired by a prominent New York attorney and having among its membership a prominent physician from the faculty at Johns Hopkins. At the 1913 meeting of the association, the report was delivered (Laughlin 1914) and read in part:

> Biologists tell us that whether of wholly defective inheritance or because of an insurmountable tendency toward defect, which is innate, members of the following classes must generally be considered as socially unfit and their supply should if possible be eliminated from the human stock if we would maintain or raise the level of quality essentials to the progress of the nation and our race:
> The Feeble Minded, The Pauper class (pauper families through successive generations); Criminaloids (persons born with marked criminal tendencies); Epileptics, The Insane (excepting certain forms of acute insanity showing no hereditary taint); The Constitutionally Weak, or asthenic class; those predisposed to specific diseases or the diathetic class; the Congenitally Deformed, and those having defective sense organs, such as the deaf-mutes, the deaf and the blind...

California had one of the longest-running involuntary sterilization programs in the country. In 1927, a team of prominent and respected citizens were assembled to consult on the effectiveness of this program. They included Lewis Terman, the most prominent psychometrician in the country, David Starr Jordan (president at Stanford), and S. J. Holmes, a distinguished geneticist from Berkeley. Covering the period from 1909 to 1927, a series of reports came out of this group that produced "the first comprehensive 'proof' that sterilization was cost-effective and posed no significant medical harm to the institutionalized persons at whom it was aimed" (Reilly 1991).

In the 1920s, two major legal developments shaped and were shaped in turn by increasingly dominant views of "race betterment" through biological science and its applications. The first was the new immigration laws, strongly backed by the old American stock, to close off the immigration doors to those from Southern and Eastern Europe. It is well known and fully documented how geneticists and eugenicists provided vital testimony before the U.S. Congress in the early 1920s (see Haller 1963; Ludmerer 1972). This resulted in the passage of two laws, which in effect reduced immigration from a surging high of approximately half a million a year to less than about 10,000 by the end of that decade. This strategy was overwhelmingly popular among local politicians in several key states. For example, the Virginia legislature passed an involuntary sterilization bill (30–0 in the senate, 75–2 in the house), noting that "heredity plays an important part in the transmission of insanity, idiocy, imbecility, epilepsy, and crime." That law gave the power to superintendents of five state institutions to petition for permission to sterilize selected inmates. This is the backdrop for the second major legal development, the famous Supreme Court decision, *Buck v. Bell*. Virginia immediately moved to sterilize an 18-year-old, Carrie Buck, the pregnant and illegitimate daughter of an allegedly feeble-minded woman.

On May 2, 1927, the Supreme Court upheld Virginia's involuntary sterilization law, opening up not only a floodgate of sterilizations in the United States, but also a model that would soon be adopted, expanded, and forever made infamous by Hitler's Third Reich. In mid-July 1933, Germany enacted a eugenic sterilization law. The American eugenicists provided the intellectual and ideological underpinnings and were widely cited as the genetic authorities in behalf of this development. California was one of the leading states in the country in terms of its use of involuntary sterilization laws. From 1930 to 1944, over 11,000 Californians were sterilized under these laws. The Germans cited the California development as a model (in 1936, Heidelberg University awarded honorary degrees to several key American eugenicists), but they took it much further. In the first year of the German program, 52,000 were placed under final order to be sterilized, a development that was in turn hailed by American eugenicists. In the period from 1933 to 1945, the best estimates indicate that the Nazis sterilized approximately 3,500,000 people (Reilly 1991).

With hindsight, we can see that there was a false precision to their connection between genetics and social troubles. As noted above, the most celebrated case of all is the sterilization of Carrie Buck. Yet, no less a learned and respected personage than Supreme Court Justice Oliver Wendell Holmes would speak for the understanding of a nation when he pronounced that Ms. Buck could and should be sterilized because "three generations of imbeciles is enough" (Smith and Nelson 1989). (It is one of the dark ironies of our national history that the daughter that Carrie Buck bore turned out to be normal.) Today, the United States is heading down a road of parallel false precision in this faith in the connection between genes and social outcomes. This is being played out on a stage with converging preoccupations and tangled webs that interlace crime, race, and genetic explanations.

So-called "genetic studies" of criminality have a heavy dependency on incarcerated populations. One of the more controversial issues in the "genetics" of crime is whether males with the extra Y chromosome, or XYY males, are more likely to be found in prisons than are XY males. The first major study suggesting a genetic link came from Edinburgh, Scotland. All of the 197 males in this account of prison hospital inmates were described as "dangerously violent" (Jacobs et al. 1965), but seven had the XYY karyotype. These seven males constituted about 3.5% of the total. Since it is estimated that only about 1.3% of all males have the XYY chromosomal makeup, the authors posited that the extra Y significantly increased one's chances of being incarcerated. Ever since, a controversy has raged as to the meaning of these findings and the methodology that produced them. Notice the logic of such studies: They argue the genetic link to crime, but rely primarily on incarcerated populations. Yet, incarceration rates are a function of incarceration decisions, a fact which social science research has long shown to be a function of social, economic, and political factors (Skolnick 1966; Currie 1985).

My point here is not to critique these studies, but to point to the dangerous precedent of this methodology in producing spurious "findings" with regard to any comparative work on genetics, crime, and race. If the method(o)logic is to compare some genetic attribute (e.g., the phenotype) of the incarcerated with the distribution outside prison walls, a host of "precise" correlations can easily be made that may have little to do with genetic causation.

During the last decade, America has been building more prisons and incarcerating more people than at any other time in our history. Indeed, in the brief period from 1981 to 1991, we went from a prison population of 330,000 inmates in state and federal prisons to 804,000. That rate constitutes substantially more than a doubling in a single decade, *the greatest rise in a prison population in modern history.*[1] (This does not include jail populations, persons in drug-related facilities, or juveniles.) Federal drug offense convictions went up 213% in this same period, signaling the importance of the "Drug War" in this development.

Converging on this development is the racial patterning of those arrested and serving time. African Americans are incarcerated at a rate approximately 7 times greater than that of Americans of European descent. Here is the beginning of the first unraveling of a false precision. Although the current incarceration rate of African Americans in American prisons is approximately 7 times that of white Americans, this is a very recent development. If we are ignorant of recent history, and do not know that the incarceration rate and the coloring of our prisons is a function of dramatic changes in the last half century, we are far more vulnerable to the seduction of a genetic explanation.[2] Table 1 reveals an astonishing pattern of prison incarceration rates by race that should give pause to anyone who would try to explain these developments by reference to the genetic makeup of the incarcerated. The gene pool among humans takes many centuries to change, but notice that since 1933, the incarceration rates of African Americans in relation to whites has gone up in a striking manner. In 1933, blacks were incarcerated at a rate approximately three times the rate of incarceration for whites (see Table 1). In 1950, the ratio had in-

---

[1] Bureau of Justice Statistics, U.S. Department of Justice, Office of Justice Programs, January 1992, Vol. 1., No. 3, NCJ-133097, Washington, D.C.

[2] For example, in 1954, black and white youth unemployment in American was equal, with blacks actually having a slightly higher rate of employment in the age group 16–19. By 1984, the black unemployment rate had nearly quadrupled, whereas the white rate had increased only marginally.

**Table 1**
Incarceration Rates by Race

| Year | Population[a] | | | Incarceration[b] | | | Rate (%)[c] | | |
|---|---|---|---|---|---|---|---|---|---|
| | total | white | black | total | white | black | total | white | black |
| 1933 | 125,579 | 112,815 | 12,764 | 137,997 | 102,118 | 31,739 | 0.11 | 0.09 | 0.25 |
| 1950 | 151,684 | 135,814 | 15,870 | 178,065 | 115,742 | 60,542 | 0.12 | 0.09 | 0.38 |
| 1960 | 180,671 | 160,023 | 19,006 | 226,065 | 138,070 | 83,747 | 0.13 | 0.09 | 0.44 |
| 1970 | 204,879 | 179,491 | 22,787 | 198,831 | 115,322 | 81,520 | 0.10 | 0.06 | 0.36 |
| 1989 | 248,240 | 208,961 | 30,660 | 712,563 | 343,550 | 334,952 | 0.29 | 0.16 | 1.09 |

[a]Total population of the United States by ethnicity (in thousands). Source: Series A 23–28: Annual estimates of the Population, by Sex and Race: 1900 to 1970. In *Historical Statistics of the United States, 1976*: Department of Commerce, Bureau of the Census, 1976, pp. 9–28. No. 19: Resident Population—Selected Characteristics: 1790–1989. *Statistical Abstract of the United States 1991*, 111th edition, Bureau of the Census, 1991, p. 17.
[b]Total number of prison population by ethnicity. Note: Data for incarceration reflect the estimated number of prisoners surveyed on a particular date. Source: Table 3-31 Characteristics of Persons in State and Federal Prisons. *Historical Corrections Statistics in the United States*, Bureau of Prison Statistics, 1986, p. 65.
[c]Incarceration/population.

creased to approximately 4 to 1; in 1960, it was 5 to 1; in 1970, it was 6 to 1; and in 1989, it was 7 to 1.

Even in the last decade there has been a sharply focused and dramatic issue with respect to incarceration rates by race. In Florida, the annual admissions rate to the state prison system nearly tripled from 1983 to 1989, from 14,301 to nearly 40,000 (Austin and McVey 1989). This was a direct consequence of the "War on Drugs," since well over two-thirds of these felonies are drug-related. The drug war affected the races quite differently with regard to respective prison incarceration rates. The most astonishing figure is for the state of Virginia: In 1983, approximately 63% of the new prison commitments for drugs were white, with the remaining 37% minority. Just 6 years later, in 1989, the situation had reversed, with only 34% of the new drug commitments being whites, but 65% minority. Yet, in this very period we find a significant increase in scientific journal articles and scholarly books (Mednick et al. 1984; Wilson and Herrnstein 1985) suggesting a greater role for biological explanations of crime.

In a parallel fashion, today's celebrants of the new advances in human genetics conjure up a future of the elimination of genetic disorders, either by prevention of the births of those who have such disorders, or by the promissory note of treatment. Yet, although we can detect and diagnose hundreds of genetic disorders, we have effective treatments for almost none (Nichols 1988). Thus, it is to prevention that we are bound to turn, and with that turn there is nothing to restrain the helmsman from the eugenic shores.

The importance of this turn toward prevention for the impending conversion of genetics, race, and crime is particularly ominous. If we take the three together, genetics, race, and crime, we should be struck by a remarkable thing—two of the three lack anything resembling precision. A strong motive force behind the science of genetics is that, if we can just get down to the molecular genetic level, we can have greater precision and control. Yet, what I am about to argue here is that the "outcome variables" of crime and race are such as to render any such "relationship" spurious. Genetic explanations are based on the assumption that there is a finite expression of the gene, as in whether one will be nearsighted, have certain hair texture, etc., and perhaps ultimately in behaviors, competencies, attitudes, health vulnerabilities, even desires. Whether one will develop lung cancer in combination with other factors such as smoking or inhaling toxic fumes for many years is of course a combination of the genetic and behavioral. Even such circumscribed outcomes within humans as cancer and heart disease are rarely determined by single genes. The vast majority of human disorders and diseases can best be explained by considering complex multifactorial environmentally influenced interactions. However, for the sake of argument, let us proceed with the assumption that the gene is a finite, relatively precise thing with predictable outcomes that are not tied up with thousands of contingencies and variables.

What constitutes either a "crime" or a "race" is systematically patterned, yet variably determined. This is demonstrable as one moves from one nation to another, sometimes from one region to another, and even from one decade to another in the same society. Let us consider crime.

As noted above, studies that have attempted to claim a genetic component to crime are almost always confined to an examination of existing *institutional* records.[3] The methodological flaw revolves around this limitation. There are three crucial reasons that studies of crime and genetics are so problematic that it is almost impossible to have much confidence in them. First, the very definition of what constitutes a crime is highly socially variable, depending on the passage of a law, the policing practices and the judicial system of a society, and a particular point in history. Just breaking into a factory to steal linen was once deemed a crime so heinous as to require the death penalty:

> In 1764 Parliament decreed that the death penalty would apply to those who broke into buildings to steal or destroy linen, or the tools to make it, or to cut it into bleaching-grounds. But the penalties were contained in an incidental clause in an act passed to incorporate the English Linen Company, whose proprietors included Lord Verney and the Right Honorable Charles Townsend. (Hay et al. 1975)

The limitations of looking only at institutional records becomes apparent if one looks at the record of rape in the southern United States. Over a 250-year history, at least if one were relying on "incarceration rates," one would come to the conclusion that no white man ever committed the crime of rape on a black woman in 12 southern states. If a Scandinavian came to the United States to study crime and genetics and the research was based on *records*, this whole population of "criminals" would be missed.[4]

Second, the term "criminal" lumps together the one-time offender with the career criminal, the professional isolated con artist with a bureaucrat in organized crime; it lumps together the hit-and-run driver with the rapist; it even lumps together the inadvertent poacher on the land of the gentry with a deliberately adulterous member of the gentry (Hay et al. 1975). What it typically does not do is lump together the crimes of corporate executives with the crimes of the common thief and burglar. The implications of this did not escape the notice of the Supreme Court of the United States. In *Skinner v. Oklahoma*, the Supreme Court ruled in

---

[3]Rowe (1986) does use a self-report mailed questionnaire for data for a twin study. In this article, his focus is on "anti-social behavior."

[4]Those who would suggest that rape during slavery is a political concept are on a slippery slope of logic that leads not only to joining with an old Eldridge Cleaver formulation about rape during rebellious periods as politics, but even more radically, to then have to address a rhetorical question from the Marxists: When is crime not political?

1942 that the sterilization of a man because he was a third-generation criminal was in violation of the equal protection clause of the constitution. The State of Oklahoma's *Habitual Criminal Sterilization Act* had provided that one convicted of three felonies could be "rendered sexually sterile."

> Skinner had been convicted of stealing chickens in 1926 and had been found guilty of armed robbery in 1926 and 1929... (the U.S. Supreme Court) chose to overturn the law because it violated the Fourteenth Amendment's guarantee of the equal protection of the laws. Oklahoma had exempted certain kinds of felons from its reach. Offenses against prohibition, revenue laws, embezzlement, and political crimes were deemed insufficient for sterilization. This clear bias in favor of white collar criminals was judged by the Supreme Court to be "unmistakable discrimination." (Reilly 1977)

The Supreme Court ruled, however, that such sterilization was *unwarranted because the prosecution had not demonstrated that the more privileged classes could not similarly be prosecuted.*

Finally, what constitutes a criminal is far more arbitrarily defined in terms of an empirical referent for research than it appears on the surface. One way of defining a criminal is to simply say that it is someone who has committed a crime. However, there is a competing definition, which characterizes a criminal as someone who has been *convicted* of a crime. Not all those who commit crimes are convicted.[5] Only a small percentage of such persons are arrested, fewer still are prosecuted, only a fraction of these are convicted, and even a smaller percentage are incarcerated. For the bulk of crimes committed (reported and known to the police), the fall-away rate can be as high as 80%. In Skolnick's (1966) earlier study of a police department in a major city of the United States, less than 25% of the burglaries were cleared by arrest and prosecution (much less conviction and incarceration). The figure was only slightly higher for robberies, a repeated pattern in reports from around the nation (Skolnick 1967), and this has not changed in the last two decades (Federal Bureau of Investigation 1988).

If one were doing a *correlational* study looking for the "genetic component" in imprisonment, one could find a "significant" statistic that would show the importance of "genes" (in this case, race). *If one were to use the data presented above on incarceration rates with correlational studies, by race, one would have far more significant correlations than any previously reported in all the twin studies of IQ and genetics; crime, kinship, and genetics; or in any of the adoptive/biological*

---

[5]It has been estimated that Americans steal $38,000,000 per day in shoplifting alone. The President's Crime Commission's survey of 10,000 household concluded that "91% of all Americans have violated laws that could have subjected them to a term of imprisonment at one time in their lives." (Reiman 1984, p. 81)

*relative studies of schizophrenia.* If one were looking for "genetic" evidence, using the same theoretical approach and methodological techniques of reviewing institutional records, such a correlational study would conclude that there is a high probability of a "criminal gene" and that it is related to "race." Since there is remarkable variability from one time to another as to what constitutes a serious crime, there can be high variability in which "race" commits those crimes. We know in our own history that horse thievery was a greater crime than murder, well into the nineteenth century. Of course, selling silk stockings in Moscow was a crime in 1978, but a boon in 1992; and shooting fleeing East Germans was a crime in July of 1991, but the source of military affirmation two years earlier. The examples are endless, and the point is clear. Well, if crime is so problematic across a short time frame and a short space, what about race?

Just 80 years ago, it was common to refer to the Slavic Race, the Jewish Race, and yes, the Aryan Race. For a time, "science in the form of physical anthropology" took over this "ethnic" classification of race and played a major role in reformulating the more current broad categories of Caucasoid, Negroid, and Mongoloid. This classification scheme makes perfect sense only if race is appreciated as a social category. During more than two centuries of slavery in the United States, there was considerable mixing between whites and blacks, but the offspring of all such unions, where socially traceable, were considered black. Moreover, that legacy remains today, accounting for the wide phenotypic variation among African Americans. Yet, studies of "race" and a wide variety of other matters, such as IQ, crime, and hypertension have been attempted with genetic explanations.

Even with strong epidemiological evidence that heart disease and hypertension among African Americans is strongly associated with such social factors as poverty, there has been a persistent attempt to pursue the scientific study of hypertension through a link to the genetics of race. Dark pigmentation is indeed associated with hypertension in America. Michael Klag and his colleagues recently reported the results of a carefully controlled study looking at the relationship between skin color and high blood pressure (Whittle 1991). They found that darker skin color is a good predictor of hypertension among blacks of low socioeconomic status, but not for blacks of any shade who are "well employed or better educated." The study further suggested that poor blacks with darker skin color experience greater rates of hypertension not for genetic reasons but because darker skin color subjects them to greater discrimination, with consequently greater stress and psychological/medical consequences.

The complexity of the interaction between genetic and environmental factors is a given, but it is *how* they interact and the relative weights assigned to each that is the source of contention in the nature/nurture debates. Given the dramatically increasing coloration of America's prison population, a proposed study of the relationship between skin color and incarceration rates is probably not far away. If this study were done cross-sectionally (at a particular time, across a range of prisons) as opposed to longitudinally (tracing patterns of imprisonment by race over a long period of time, e.g., several decades), one might conclude that the genetic explanation of race differences in incarceration was more compelling. Yet all we can say with any certainty in a cross-sectional analysis is that the rate of incarceration is so much higher for blacks than whites. Such spurious correlations are the source of false precision. It is not much of a reach to suggest that the new technologies using DNA identification will provide part of the larger context in which the imprimatur of legitimacy may yet enshrine "genetic thinking" about crime. Like Ulysses, we had better strap ourselves to the masts in preparation for a spate of such research, which has languished only because the political climate for it has been inhospitable. That could easily change.

## REFERENCES

Austin, J.S. and A.D. McVey. 1989. The impact of the war on drugs. The 1989 National Council of Crime and Delinquency Prison Population Forecast, San Francisco, California. *Focus* **39**: 1.

Currie, E. 1985. *Confronting crime: An American challenge.* Pantheon, New York.

Federal Bureau of Investigation. 1988. *Uniform crime reports for the United States.* Department of Justice, Washington, D.C.

Gould, S.J. 1985. *The flamingo's smile: Reflections in natural history*, pp. 319–332. W.W. Norton, New York.

Haller, M.H. 1963. *Eugenics: Hereditarian attitudes in American thought.* Rutgers University Press, New Brunswick, New Jersey.

Hay, D., P. Linebaugh, J.G. Rule, E.P. Thompson, and C. Winslow. 1975. *Albion's fatal tree. Crime and society in eighteenth century England*, pp. 189–253. Pantheon, New York.

Herrnstein, R.J. 1989. IQ and falling birth rates. *The Atlantic Monthly* **255**: 73.

Jacobs, P.A., A.M. Brunton, M. Melville, R. Brittain, and W. McClemont. 1965. Aggressive behavior, mental subnormality, and the XYY male. *Nature* **208**: 1351.

Laughlin, H.H. 1914. *The Scope of the Committee's Work,* Bulletin No. 10A, pp. 12–13. Eugenics Records Office, Cold Spring Harbor, New York.

Ludmerer, K.M. 1972. *Genetics and American society.* The John Hopkins University Press, Baltimore, Maryland.

Mednick, S.A., W.F. Gabrelli, Jr., and B. Hutchins. 1984. Genetic influences in criminal convictions: Evidence from an adoption cohort. *Science* **224**: 891.

Milunsky, A. 1992. *Heredity and your family's health.* Johns Hopkins University Press, Baltimore, Maryland.

Nichols, E.K. 1988. *Human gene therapy.* Harvard University Press, Cambridge, Massachusetts.

Reilly, P. 1977. *Genetics, law, and the social policy*, pp. 126–127. Harvard University Press, Cambridge, Massachusetts.

———. 1991. *The surgical solution: A history of involuntary sterilization in the United States.* Johns Hopkins University Press, Baltimore, Maryland.

Reiman, J.H. 1984. *The rich get richer and the poor get prison.* John Wiley, New York.

Rowe, D.C. 1986. Genetic and environmental components of antisocial behavior: A study of 265 twin pairs. *Criminology* **243:** 513.

Skolnick, J. 1966. *Justice without trial: Law enforcement in a democratic society.* John Wiley, New York.

———. 1967. *Coercion to virtue: A sociological discussion of the enforcement of morals.* President's Commission on Law Enforcement and Administration of Justice, Washington, D.C.

Smith, D.J. and K.R. Nelson. 1989. *The sterilization of Carrie Buck.* New Horizon Press, Far Hills, New Jersey.

Whittle, J.C., P.K. Whelton, A.J. Seidler, and M.J. Klag. 1991. Does racial variation in risk factors explain black-white differences in the incidence of hypertensive end-stage renal disease? *Arch. Int. Med.* **151:** 1359.

Wilson, J.Q. and R. Herrnstein. 1985. *Crime and human nature.* Simon and Schuster, New York.

# DNA Data Banking and the Public Interest

NACHAMA L. WILKER,[1] STEVEN STAWSKI,[1] RICHARD LEWONTIN,[2] AND PAUL R. BILLINGS[3]

[1]Council for Responsible Genetics
Cambridge, Massachusetts 02138
[2]Harvard University
Cambridge, Massachusetts 02138
[3]Division of Genetic Medicine
California Pacific Medical Center
San Francisco, California 94115

## INTRODUCTION

This chapter presents the current thinking of the Human Genetics Committee of the Council for Responsible Genetics (CRG) on issues related to the storage and use of genetic information. We briefly discuss technical problems with DNA-based identification systems as they have been applied in the courts. However, since many of the problems with DNA profiling methods may be resolved within the next 10 years, the main focus of this chapter is on the collection, storage, and use of DNA information in data banks. Specifically, we outline what types of information are stored and discuss the current and future impact of the gathering and use of personal genetic information.

As a national organization of scientists and other interested persons, the CRG is dedicated to promoting discussion about new genetic technologies. The CRG's mission is to support biotechnology in the public interest while alerting the public to the social and environmental problems arising from new developments in human genetics. Through its programs, the CRG hopes to bring new genetic technologies under public control.

## DNA PROFILING IN THE CRIMINAL CONTEXT

On July 22, 1992, the Supreme Judicial Court of Massachusetts (*Commonwealth v. Lanigan* 413 Mass. 154) ruled that the results of a DNA identification test were inadmissible due to unresolved technical problems with the test. One month later, the California Court of Appeals followed suit (*People v. Barney* 92 C.D.O.S. 6851). The judgments of both courts were based on technical questions about how the samples were handled and theoretical concerns surrounding how the results were interpreted. These decisions highlight conflicts within the scientific community's view of the decisiveness of these tests, particularly in determining whether a suspect might have committed a crime (National Research Council 1992).

Forensic DNA identification tests usually compare DNA extracted from blood, hair, or other tissue samples recovered at the scene of a crime to DNA taken from the tissues of a suspect or victim. After analysis, a laboratory representative testifies in court as to whether the suspect's or victim's DNA profile "matches" the sample left at the crime scene *and* the likelihood that such a matching profile could have been provided by an individual other than the suspect. Technical inaccuracies can arise with contamination- and degradation-induced inaccuracies, amplification oversensitivity, and artifacts induced during the gel analysis. However, this method's greatest technical obstacles arise in the calculation of the probability of coincidental matches and the criteria used for determining the "relevant" population.

For a "match" to occur, the DNA identification method must include a determination of how likely it is that a person, other than the suspect or the victim, might have donated the sample found at the scene of the crime. This statistic is calculated by comparing the observed profile to a reference population. Herein lies the problem; there are no standards for constructing or identifying appropriate reference populations (Lewontin and Hartl 1991). A case in point is that of the *State of Vermont v. Passino*, in which Passino was accused of the assault and murder of a woman. A match was made between the suspect's DNA profile and DNA found at the scene of the crime. Passino, an Abenaki Indian, disputed the DNA evidence on the basis that no database exists that accurately reflects his ethnic background.

The crime occurred in rural Franklin County, Vermont, which is located on the border with Quebec and has the highest proportion of Indians, largely Abenaki, of any county in the state. The Abenaki and Abenaki/French Canadian mixed families comprise a population that straddles the United States and Canadian border. They reside predominantly in trailer parks and are chronically under- and unemployed, impoverished, and without social supports and entitlements. The victim, who was herself half Abenaki, was assaulted and killed in a trailer camp where she lived and in which a large fraction of the other residents were also of Abenaki ancestry. Given the circumstances of the crime, the defense could make the entirely reasonable claim that it is indeed the Abenaki who comprise "the population" of potential suspects, of whom the defendant is only one. On the other hand, the prosecution might reasonably claim that a trailer camp on a state highway is accessible to any passing motorist, so the entire population of western Vermont and eastern New York is the appropriate reference group. Rather than determining the appropriate reference population by means of objective science, we are faced with subjective arguments about the patterns of people's lives.

Yet despite the shortcomings of DNA profiling, the debate over its appropriateness and utility is unlikely to end. Although the method poses significant unresolved questions in its forensic applications, it will likely remain a part of criminal investigations and the adjudication process. By the end of this decade, advances in sequencing techniques will permit the direct comparison of DNA samples without recourse to reference populations. At that point, the only remaining uncertainties will involve the technical competence of the laboratories performing the profiles. The need for oversight, rigorous technical standards, and supervision of these laboratories will remain.

## WHAT DATA BANKS ARE BEING ESTABLISHED?

Currently, the Federal Bureau of Investigation (FBI), insurance companies, state governments, and the armed services have begun to collect and store genetic information in data banks. Outside the criminal context, such data banks include information gathered from genetic tests conducted to determine paternity and disease susceptibility, and from medical records or physician's invoices that have been submitted to insurers, state or federal agencies for payment.

Once personal medical or genetic information is obtained and stored in data banks by government agencies or private companies, the potential for violations of civil liberties expands dramatically. Important genetic and other personal information can be extracted from an individual's blood, including data about the presence of disease-associated genes, medical conditions, or current drug use. In data banks that store biological samples, such as tissues or blood, the sample can be analyzed for genetic information that reveals potential health risks and other private information. The unauthorized testing of materials and use of data present a grave threat to civil liberties; there are few preventing misuse of data banks.

For the purpose of this paper, the variety of existing data banks can be loosely categorized into *identification* and *information* banks. Identification banks contain computer records of physical characteristics (height, weight, hair and eye color, skin marks, dental variations, fingerprint patterns) and are designed to be used by businesses and government agencies to identify individuals. Information banks, on the other hand, contain more detailed personal information (including medical records, credit or police reports, billing records) and are used primarily by hospitals, blood laboratories, transplant services, employers, and insurance companies to obtain or store detailed information on specific individuals. Some identification and information banks maintain tissues (such as kidneys and blood) and biological samples (including pathological specimens from surgeries or biopsies, amniocentesis samples, and Guthrie blood spots). Both identification and information banks are being expanded to include genetic information.

### Identification Banks

More than 18 states, the FBI, and the Department of Defense (DOD) are expanding identification data bank systems to store genetic information. Although the primary intent of public officials may be to use computerized DNA profiles to identify criminals or missing persons, data banks that store biological samples (and not just computerized profiles) permit the misuse of personal genetic information. There may be unanticipated information in the simple DNA profile patterns. As the techniques improve and more information can be extracted, misuse becomes inevitable.

Virginia officials, for example, have collected blood and saliva samples from more than 40,000 convicted felons, making their forensic data bank the largest in the nation (Keehn 1992). Authorities plan to compare DNA profiles of criminals with those obtained from blood and semen samples found at sites of murder, rape, and other violent crimes. Virginia's legislation, however, requires the collection of biological samples from convicted felons—including tax evaders and burglars—who are unlikely to leave biological samples at crime sites or to repeat their offenses. By including these individuals in its database, Virginia is expanding collection and storage of DNA information to a portion of the population that is not relevant to the stated purpose of criminal investigations.

The FBI is also establishing a centralized identification database that contains information from DNA analysis of violent offenders (Brewin 1991). Results from the National DNA Identification Index (NDII), the FBI's pilot program, are aiding in the development of relevant software and its distribution to state, county, and city forensic laboratories. The intent of this computer network is to make DNA identification information centrally available. Its effectiveness and potential harms are still uncertain. It is, however, worrisome that the FBI has resisted attempts at oversight and regulation of its practices (see Kotval, this volume).[1]

A third example of an identification bank containing genetic information is the DOD's project to identify possible future war casualties. In April of 1992, the department began to collect biological samples from armed services personnel—eventually some two million people—in order to compare their DNA profiles with those from unidentifiable, "unknown soldiers" (Leary 1992; Recer 1992). The DOD stores duplicate biological samples which could later be used to extract health and other information. Routine access to this data bank by other federal agencies will not be allowed in theory; in practice, it is likely that once the information is collected and banked, pressures will mount to use it for other purposes.

### Information Banks

Unlike identification banks, which are intended solely for identification purposes, information banks are designed to store detailed personal information. For clarity,

---

[1]*Editor's note:* The FBI program does not exclude the use and storage of information (other than DNA profiling patterns) that may arise from biological samples stored by state or local agencies.

we classify information banks into three general types and consider them separately below:

1. *Information data banks that store identification items (names, addresses, etc.) and detailed personal information including medical, police, and credit reports, in computers.* The data bank maintained by the Medical Information Bureau Inc. (MIB) in Westwood, Massachusetts, for example, contains medical information on 10–20 million people in all parts of the country. The data bank was designed to lessen the incidence of insurance fraud, but it is now used by insurance companies to set an applicant's premiums or to assist in decisions about offering coverage. This information allows insurers to identify potential clients who have tested positive for certain genetic disorders (like Huntington's disease or acute intermittent porphyria) and to exclude these conditions from coverage or refuse to offer contracts to these individuals.

   Most people are unaware that their personal genetic or medical information is stored in MIB's data banks and used for insurance business purposes. Although the establishment of an MIB file or the initiation of a data bank search requires "consent," failure to agree to MIB searches is grounds for exclusion from insurance coverage.

2. *Information data banks that store biological samples including pathological specimens from surgeries or biopsies, amniocentesis samples, and Guthrie blood spots and are used primarily by hospitals and laboratories.* Advances in the technical ability to extract genetic information from preserved or exceptionally small biological samples pose dangers to privacy because these samples potentially contain vast amounts of personal information.

   The small sample of a newborn's blood from the card taken for the Guthrie test, for example, can (even after years of storage) yield enough DNA for hundreds of genetic tests (Nelson et al. 1990; Schwartz et al. 1990; Matsubara et al. 1991). Although some of the materials subsequently stored in tissue banks now undergo limited genetic analysis (karyotyping of amniocentesis samples and biochemical genetic testing of Guthrie spots), minor methodological advances would enable technicians to perform hundreds or thousands of genetic tests on these samples (Billings 1992). Neonatal medical screening is routinely done to collect biological samples, and informed consent by families does not always accompany the screening. Because many states do not have laws regulating the storage or use of these materials, tissue banks could readily become vehicles for population-wide genetic testing that could produce serious invasions of privacy.[2]

3. *Information data banks that store both computerized information and biologi-*

---

[2]*Editor's note:* The remarkable paper of Matsubara et al. (1991) clearly illustrates the dangers that these data banks pose to the right of privacy and to be informed about medical testing.

*cal samples.* These banks are used by hospitals, laboratories, blood banks, and transplant services to record computer-based characterizations of genetic tissue traits (blood groups, histocompatibility antigens, serological reactions, etc.) and are the most common form of data bank in current use. Some services keep tissue collected from individuals (donated units of blood or kidneys) along with other computer-stored information. These data banks also have the potential to be misused in nonconsensual ways and to violate an individual's privacy.

## THE IMPACT OF DNA DATA BANKING

Although the routine storage of genetic information within large data banks is already under way, there has been little public discussion of the problems or threats to civil liberties such programs pose. As noted previously, not only can tissue or DNA samples identify individuals, but they can also be used to produce information related to health, paternity, and other personal issues.

Proponents of the use of DNA-based identification systems and data banks point to their utilization in situations that arouse public sympathy, such as tracking down serial murderers, finding missing children, reuniting broken families, and controlling sex offenders. In reality, however, the net of mandatory sampling and testing inevitably will include situations that were not contemplated originally. In the 1970s, when mandatory sickle cell screening was attempted, the results of tests that identified those individuals who had the sickle cell trait, which involves no symptoms, were used in an adverse discriminatory manner by insurance companies and the armed services (Duster 1990).

Serious civil liberty concerns are raised when databases are created on the basis of nonconsensual sampling or analysis of DNA. As Americans, we have prided ourselves on living in a country where we are not registered or otherwise identifiable by our government. Over the years, this tradition of anonymity has been eroded by routine use of Social Security numbers in contexts that have no bearing on the Social Security program. Similarly, fingerprints which were introduced to identify accused felons are now used by schools, employers, professional associations, and other private and government organizations for purposes unrelated to criminal prosecutions.

The use of genetic information within data banks by insurers, employers, governmental agencies, and other parties has already resulted in genetic discrimination and led to invidious social stratification (Lippman et al. 1990). Access to health care, financial security, home ownership, employment, and the freedom to change jobs and improve working conditions are becoming limited to individuals with acceptable "genetic profiles" (Billings et al. 1992). The right of individuals to "know" or "not know" information about themselves, while retaining the full measure of individual liberties, is threatened when important social, business, and governmental institutions are permitted to institute genetic testing and data banking programs.

Finally, another problem may arise if researchers try to use biological samples collected and stored in forensic contexts in experiments designed to identify genes associated with criminal behavior. Although such research has no scientific merit or basis, it could be used as a new biological justification to bolster racist and ethnic prejudice. The continued popularity of behavioral genetics makes it quite possible and even likely that selected population data banks will be misused in such "research" (Butterfield 1992).

## POLICY CONSIDERATIONS

In light of the problems associated with DNA data banking, the CRG believes that the public must demand an accounting and justification for the allocation of resources to these practices by local, state, and federal agencies. In addition, the CRG calls for certain rules to be applied to the collection, storage, and use of genetic information:

1. Genetic information gathered for prenatal, neonatal, transplantation, blood typing, identification, and other screening purposes by private or public agencies must be restricted to the specific purpose for which it was collected.
2. Until safeguards are in place to prevent misuse of genetic information or stored biological samples, data banks designed for identification purposes must not contain biological samples and should only be used with the donor's permission. When this is not possible, an impartial judicial review should control access to this information, and a "necessity" or "public good" standard should be met before allowing any nonconsensual access.
3. Data banks must be open to public scrutiny. The techniques by which samples and data are obtained must be monitored, and the accuracy of the information must be validated. Data must not be gathered by coercion unless legally certified evidence is available that these data will aid in the identification of a violent repeat offender (judicial review standard mentioned above). Private and public agencies and companies, which already have DNA information in their files, must notify the individuals involved.
4. The storage of biological samples in data banks must not be routinely undertaken by any agency, except those with a medical purpose, and then only when consent, confidentiality, and security are well established. Donors should understand the data stored in data banks about them and should provide specific consents for each use of this information. In general, the right of governments to access and store any form of genetic information should be severely limited.
5. Government data banks that store genetic information or biological samples

should not be used for insurance-related inquiries, employment decisions, or external agency review. Specific laws must prohibit the use of data banks within branches of government and the sharing of such information between them. For instance, the FBI should be forbidden from using the DOD's genetic data bank.

The CRG believes that there is a pressing need for the development of strict rules to govern the banking of information or biological samples. Only when the proliferation of data development and storage is curbed, and the rights and entitlements of individuals are reaffirmed, will the serious dangers of genetic data banking be reduced and the potential benefits of genetic testing be realized.

## ACKNOWLEDGEMENTS

This chapter reflects discussions over the past 2 years of the Human Genetics Committee of the CRG. We thank its other members, including A. Lippman, P. Bereano, C. Gracey, M.S. Henifin, R. Hubbard, S. Krimsky, K. Messing, S. Newman, J. Norsigian, and M. Saxton. We also thank R. Hubbard, J. Glaubman, and K. Chenausky for their assistance in preparing and editing this chapter. Further information about the work of the CRG can be obtained by contacting us at 19 Garden Street, Cambridge, Massachusetts 02138 (telephone: 617-868-0807). All errors are, of course, our own.

## REFERENCES

Billings, P.R. 1992. The scientific basis of the "genetic revolution": A selective review. In *The genome, ethics and the law: Issues in genetic testing* (ed. AAAS-ABA National Conference of Lawyers and Scientists), p. 23. American Association for the Advancement of Science, Washington, D.C.

Billings, P.R., M. Kohn, M. de Cuevas, J. Beckwith, J.S. Alper, and M. Natowicz. 1992. Discrimination as a consequence of genetic testing. *Am. J. Hum. Genet.* **50:** 476.

Brewin, B. 1991. FBI plans national DNA data base. *Federal Computer Week*, June 24, p. 37.

Butterfield, F. 1992. Studies find a family link to criminality. *The New York Times*, January 31, p. 1.

Duster, T. 1990. *Backdoor to eugenics*. Routledge, Chapman and Hall, New York.

Keehn, J. 1992. The long arm of the gene. *American Way*, March 15, p. 36

Hoeffel, J.C. 1990. The dark side of DNA profiling: Unreliable scientific evidence meets the criminal defendants. *Standard Law Rev.* **42:** 465.

Leary, W.E. 1992. Genetic record to be kept on members of the military. *The New York Times* January 12, p. 15.

Lewontin, R.C. and D.L. Hartl. 1991. Population genetics in forensic DNA typing. *Science* **254:** 1745.

Lippman, A., P. Bereano, P. Billings, C. Gracey, M.S. Henifin, R. Hubbard, S. Krimsky, R.

Lewontin, K. Messing, S. Newman, J. Norsigian, M. Saxton, and N.L. Wilker. 1990. *Position paper on genetic discrimination.* Council for Responsible Genetics, Cambridge, Massachusetts.

Matsubara, Y., K. Narisawa, K. Tada, H. Ikeda, Y.Q. Yao, D.M. Danks, A. Green, and E.R. McCabe. 1991. Prevalence of K329E mutation in medium-chain acyl-coA dehydrogenase gene determined from Guthrie cards. *Lancet* **338:** 552.

National Research Council, Commission on Life Sciences, Board of Biology. 1992. *DNA technologies in forensic science.* National Academy Press, Washington, D.C.

Nelson, P.V., W.F. Carey, and C.P. Morris. 1990. Gene amplification directly from Guthrie blood spots. *Lancet* **336:** 1451.

Recer, P. 1992. U.S. to use DNA "dog tags" in casualty identification. *The Boston Globe*, January 26, p. 3.

Schwartz, E.I., S.E. Khalchitsky, R.C. Eisensmith, and S.L. Woo. 1990. Polymerase chain reaction amplification from dried blood spots on Guthrie cards. *Lancet* **336:** 639.

# Subject Index

Adversarial process, 23, 33–41, 51–56
Allelic polymorphism, 16
American legal system. *See also* Adversarial process; Presumption of innocence; Privacy; Search and seizure
  individual dignity, 30–31
  individual rights, 21–33, 51–56
  procedural truth, 23–24, 33–41
  trial by jury of one's peers, 22, 36–41

Bayes' rule, 91–96
Bertillon, Alphonse, 6–7
Bertillonage, 6–7
Bill of Rights, 130
Bin frequency, 83
Biotechnology companies, 2, 19, 33
Blood groups, 86, 91–94, 96–98
Blood typing, 3

Castro. *See* People v. Castro, 1508/87 (N.Y. Sup. Ct. 1989)
Checks and balances, 23
Civil liberties, 119–127, 143
Cobey v. State, 80 Md. App. 31, 559 A.2d 391 (1989), 43
Collins. *See* People v. Collins, 68 Cal. 2d 319, 438 P.2d 33, 66 Cal. Rptr. 497 (1968)
Commonwealth v. Curnin, 565 N.E.2d 440 (Mass. 1991), 46–47
Consumer credit information, 124
Council for Responsible Genetics (CRG), 141, 147
CRG. *See* Council for Responsible Genetics
Criminal identification systems. *See* Bertillonage; Blood typing; DNA identification methods; Fingerprints; Human leukocyte antigen (HLA) system; Voice print
Criminals, 136–138
Curnin. *See* Commonwealth v. Curnin, 565 N.E.2d 440 (Mass. 1991)

Department of Defense, 143–144
Dermatoglyphics, 2
Diphenylamine test for gunshot residue, 8
Discrimination, 125–126, 138, 146
DNA
  coding and noncoding regions, 13
  highly repetitive sequences, 2, 3
  highly variable regions, 2
  hypervariable loci, 14–15
  sequence variation among individuals, 3
DNA data banks
  compilation, 16–17, 25, 124–125, 143–147
  conflict between privacy and accuracy, 31–32
  ethnic subdivision, 88
  feasibility, 12
  and genetic privacy, 124–125
  identification banks, 143–144
  information banks, 144–146
  and the public interest, 141–148
  safeguards, 147–148
  selection of genetic markers, 13
DNA fingerprinting. *See* DNA identification methods
DNA identification methods. *See also* DNA data banks; Frye test; Restriction fragment length polymorphisms (RFLPs); Variable number of tandem repeats (VNTRs)
  access to samples, 30–33
  civil uses
    immigration, 3
    paternity lawsuits, 28–29, 56–59
    searches for missing people, 2, 144
  differentiation between populations, 14–15, 80, 84–85
  and discrimination, 125–126
  errors in laboratory procedure, 25–26
  forensic uses
    admission of DNA as evidence, 43–51
    ethical and civil liberties concerns, 115–116, 119–127, 143

151

history of DNA in trial courts, 61–76
public policy, 109–117
legal issues, 19–41
misapplication, 116, 144
probability of specific determination, 14
reference population, 106–107, 142
and rights of the accused, 51–56
standardization of procedures, 12
statistical analysis
  band shifting, 104–106
  calculation of probability, 91–93, 142
  errors in interpretation, 26–30
  likelihood ratio, 93–98
  measurement errors, 103–104
  and the outcome of trials, 56–59
  reliability, 79–88
  sampling variability, 106–107
transfer from laboratory to courtroom, 1, 20, 141–143
DNA typing. See DNA identification methods

Electrophoretic bloodstain analysis, 11
Endogamy, 84–86
Eugenics, 5, 7, 125, 130–132
Evidence
  DNA
    and adversary resource balancing, 33–36
    case studies, 65–75
    evaluation by jurors, 38–41
    probability statistics, 56–59
    rights of the accused, 51–56
    standards for admissibility, 43–51
    and the Frye test, 8–11, 36–37, 44–49, 62–64
  physical, 9
  statistical, 26–30, 46
  testimonial, 9
Expert witnesses, 53, 64–65, 74, 76

Federal Bureau of Investigation (FBI)
  and compilation of DNA data banks, 12–13, 124, 143–144
  and development of DNA identification methods, 11–12
  and the Federal Privacy Act of 1974, 123
  and problems with use of DNA evidence, 1, 111
  procedures for obtaining evidence, 9, 113
  statistical methods, 83–84, 101
  training in forensic science, 12
Federal courts, 49–51
Fetal alcohol syndrome, 126
Fingerprints
  distinctive pattern in fetus, 6
  genes for, 17
  history, 5–7, 115
  interpretation of pattern, 6
  validation of technique, 2
Five-locus match, 64, 74
Fixed-bin method, 83, 87, 101
Forensic laboratories
  accreditation program, 110, 112–115
  assessment by LEAA, 9
  errors in interpretation of statistics, 100, 103–104
  proficiency testing, 113–114
  regulation, licensing, and oversight, 109–117
Forensic science methods. See Criminal identification systems; Polygraph
Frye test
  application to DNA identification methods, 19, 36–37, 110, 112
  principles, 8–11, 62–64
  rejection of, 48–49
  three-prong, 44–47
  two-prong, 47
Frye v. United States, 293F. 1013 (D.C. Cir. 1923). See Frye test

Galton, Sir Francis, 5–7, 17
Gaussian distribution, 103
Gene flow, 84
Genetic discrimination, 125–126, 146
Genetic privacy, 111, 115–116, 122–125
Genetic screening, 126
Genetics, link with race and crime, 129–139
Guthrie blood spots, 145

Hardy-Weinberg equilibrium, 15, 87–88, 102, 107
Hartman. See State v. Hartman, 145 Wis. 2d 1, 426 N.W.2d 320 (1988)
HIV. See Human immunodeficiency virus
HLA. See Human leukocyte antigen system
Human Genome Initiative, 16–17, 111, 115, 123

Human growth hormone, 126
Human immunodeficiency virus (HIV), 123
Human leukocyte antigen (HLA) system, 3, 15–16, 28–30, 56–57
Hypervariable loci, 14–15

Identical twins, 3
Incarceration rates, 133–135
Insurance companies, 125, 143, 146
Involuntary sterilization, 130–132

Jeffries, Alec, 3, 13

King v. Tanner, 545 N.Y.S.2d 649 (N.Y. Sup. Ct. 1989), 58–59

Lander, Eric, 13–16, 26, 100
Law Enforcement Assistance Administration (LEAA), 9
LEAA. *See* Law Enforcement Assistance Administration
Lie detector test. *See* Polygraph
Likelihood ratio, 93–98, 101–104
Linkage equilibrium, 15
Lipscomb. *See* State v. Lipscomb, 574 N.E.2d 1345 (Ill. App. Ct. 1991)

Match/binning, 99, 100, 102, 104
Medical Information Bureau Inc. (MIB), 145
Mendelian inheritance, 15
MIB. *See* Medical Information Bureau Inc.
Minisatellites, 3
Multiplication rule, 84, 87

NAS. *See* National Academy of Science
National Academy of Science (NAS), 1–2, 64
National Institutes of Health (NIH), 11
New York State Forensic DNA Analysis Panel 1989, 110
NIH. *See* National Institutes of Health

Office of Technology Assessment, 2

Paraffin test. *See* Diphenylamine test for gunshot residue
Patent law, 8

Paternity lawsuits, 28–29, 56–59. *See also* State v. Hartman, 145 Wis. 2d 1, 426 N.W.2d 320 (1988)
PCR. *See* Polymerase chain reaction
Peer review process, 2
People v. Castro, 1508/87 (N.Y. Sup. Ct. 1989), 19, 26, 44–45, 99–104, 110–111
People v. Collins, 68 Cal. 2d 319, 438 P.2d 33, 66 Cal. Rptr. 497 (1968), 26–27
Peptidase A 2-1, 16
Polygraph, 8, 40, 62
Polymerase chain reaction (PCR), 66–67, 99
Population genetics, 12, 14–15, 17, 28, 80
Presumption of innocence, 22, 28–29, 81, 88
Privacy, 31–32, 110–111, 115–116, 122–125
Probability
  posterior, 92–94
  prior, 92–95
Public policy, for forensic science in New York State, 109–117

Quality control, of forensic DNA analyses, 110

Resource balancing, 33–36
Restriction fragment length polymorphisms (RFLPs), 14, 43, 61, 66, 87, 98
RFLPs. *See* Restriction fragment length polymorphisms
Rh factor, 96–98
Risk profile analysis, 13

Search and seizure, 30–31
Self-incrimination, 9
Sound spectroscopy. *See* Voice print
Standard deviations, 100, 103–104
State v. Hartman, 145 Wis. 2d 1, 426 N.W.2d 320 (1988), 28–29
State v. Lipscomb, 574 N.E.2d 1345 (Ill. App. Ct. 1991), 47
Statistics. *See also* Bayes' rule; Fixed bin method; Likelihood ratio; Probability
  bell-shaped measurement errors, 103–104
  calculation of odds, 26–27

in DNA identification, 91–108
errors in interpretation, 26–30
and the outcome of trials, 56–59
reliability, 79–88

Tay-Sachs disease, 14
Truth, 23–24, 33–41
Two Bulls. *See* United States v. Two Bulls, 918 F.2d 56 (8th Cir. 1990)

United States Constitution
Fifth Amendment, 9, 53, 81–82, 122
Fourteenth Amendment, 52–53, 81, 137
Fourth Amendment, 30–31, 122
and individual rights, 21
Ninth Amendment, 122
Sixth Amendment, 52
United States v. Two Bulls, 918 F.2d 56 (8th Cir. 1990), 49

United States v. Williams, 583 F.2d 1194 (2d Cir. 1978), 50

Variable number of tandem repeats (VNTRs), 13, 15–16, 61, 64, 67, 70, 83–84
VNTRs. *See* Variable number of tandem repeats
Voice print, 9–10, 40

Whitner v. State, 401 S.E.2d 318 (Ga. Ct. App. 1991), 54
Williams. *See* United States v. Williams, 583 F.2d 1194 (2d Cir. 1978)

XYY males, 125–126, 129, 132–133

Yelder v. State, 1991 Ala. Crim. App. LEXIS 2536, 46